Test Pilot

1,001
Things You Thought You Knew About Aviation

Also by Barry Schiff

*Flight 902 is Down!**
*The Vatican Target**
Golden Science Guide to Flying
The Boeing 707
All About Flying
Basic Meteorology
The Pilot's Digest
The Proficient Pilot, Volume 1
The Proficient Pilot, Volume 2
Flying Wisdom: The Proficient Pilot, Volume 3
Dream Aircraft

*in collaboration

Test Pilot

1,001

Things You Thought You Knew About Aviation

Barry Schiff

Foreword by Thomas B. Haines

Aviation Supplies & Academics, Inc.
Newcastle, Washington

Test Pilot: 1,001 Things You Thought You Knew About Aviation
by Barry Schiff

Aviation Supplies & Academics, Inc.
7005 132nd Place SE
Newcastle, Washington 98059-3153
Email: asa@asa2fly.com
Website: www.asa2fly.com

Published 2001 by Aviation Supplies & Academics, Inc.

Printed in the United States of America
13 12 11 10 09   9 8 7 6 5 4 3

ASA-PLT-TEST
ISBN 1-56027-425-5/978-1-56027-425-4

Photo credits: Pages 7, 65, 66, 71, 104, 123, 144, 220, 227, 237, 271, 277, 296, 305, 309, 313, 324, 330, 335, 363, 368, Barry Schiff; p. 116, Greg A. Syverson; p. 243 (partial), Henry Geijsbeek; p. 300, Mischa Hausserman; p. 385, Norman Wexler; p. 1, 41 courtesy Goodyear; p. 35 courtesy Beech Aircraft; p. 1, 58, 208, 224, 226, 229 courtesy Cessna Aircraft; p. 60-61 courtesy of National Air and Space Museum; p. 85 courtesy Patty Wagstaff; photos of Philip Dalton on pp. 123, 154 courtesy of Marcia Dalton Smith; historical tornado photo on p. 168, and p. 243 (partial), courtesy the National Oceanic and Atmospheric Administration/ Department of Commerce; portrait of St. Erasmus on p. 194, detail from Matthais Grünewald's *The Meeting of St. Erasmus and St. Maurice* (c.1520-1524, *Alte Pinakothek*, Munich, Germany, see website http://www.abcgallery.com for more information); p. 205, courtesy Bell Helicopter Textron; p. 251, 258, courtesy Boeing Commercial Airplane Group; p. 231, 243 (partial), 335, courtesy of the Lockheed Martin Corporation.

Library of Congress Cataloging-in-Publication Data:

Schiff, Barry J.
Test pilot : 1,001 things you thought you knew about aviation / by Barry Schiff ; foreword by Thomas B. Haines.
     p. cm.
     ISBN 1-56027-425-5
     1. Aeronautics—Miscellanea. 2. Airplanes—Piloting—Miscellanea.
     3. Aeronautics—Examinations, questions, etc.
TL546.5 .S35 2001
629.13—dc21   2001046409

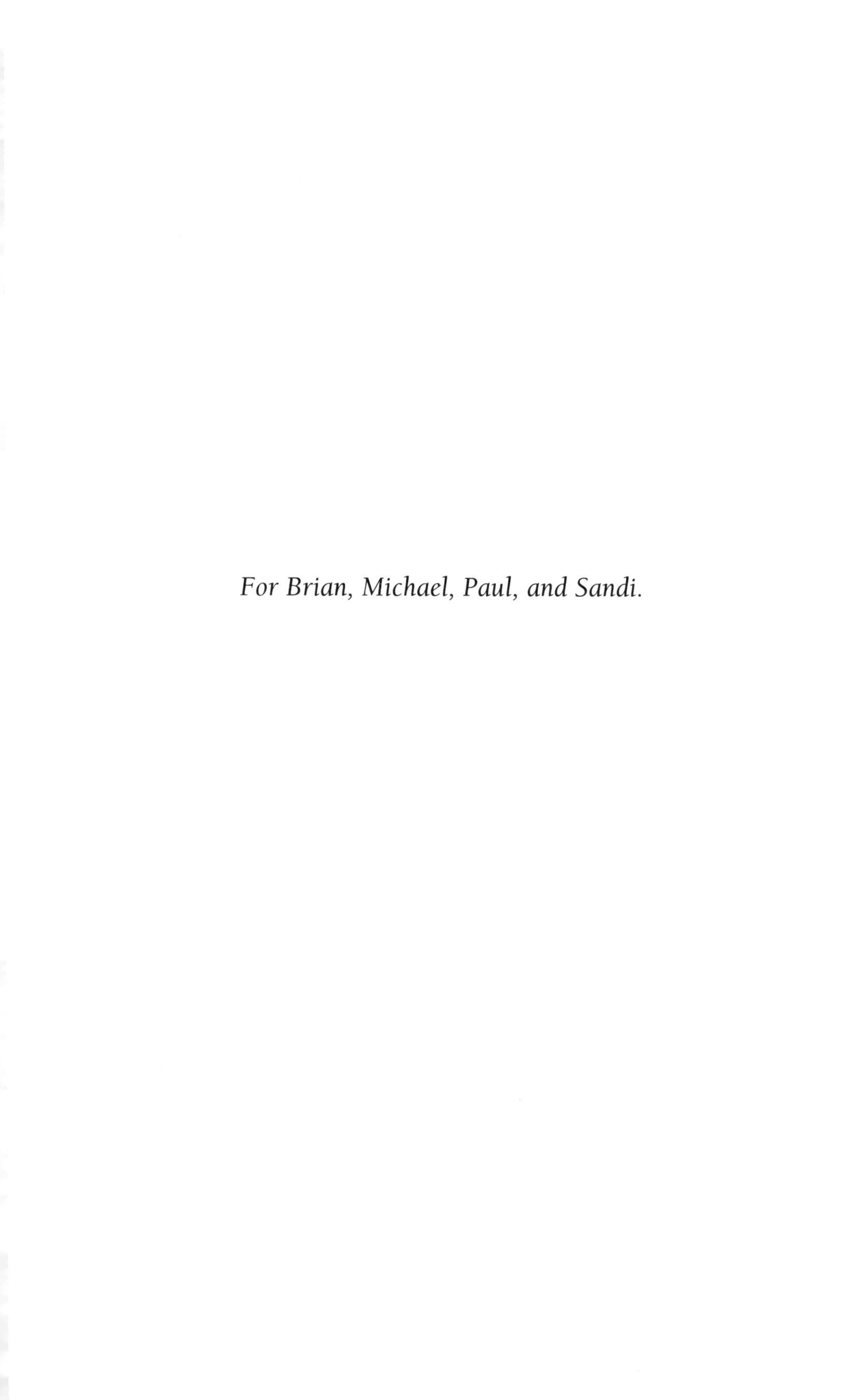

For Brian, Michael, Paul, and Sandi.

Contents

Foreword

Go to dinner with Barry Schiff and two things will happen. 1. You will get the check. 2. You will learn something about aviation.

A breath blowing over a paper napkin becomes a device for demonstrating Bernoulli's Principle. A wet finger sliding around the rim of a water glass explains the Doppler effect. Two intertwined forks, a toothpick, and a wine goblet give Barry the opportunity to teach about weight and balance and to seemingly defy gravity.

Here's a guy who lives, eats, and flies aviation. Despite more than 26,000 hours logged, a 34-year career as an airline pilot, and 40 years of writing articles for *AOPA Pilot* magazine, Barry is still fascinated by the details of aviation. Never one to take anything aeronautical for granted, he seems to be continuously on a knowledge quest. The rest of us see the word aileron and think of the flight control that causes an airplane to roll. Barry wonders where the word came from. Why is it called an aileron?

Not only does Barry want to know all of this stuff aeronautical, he also wants to share it. As a longtime flight instructor, he just can't help himself. So it was in the fall of 1993 that he began talking with us at *AOPA Pilot* about the prospects of a regular quiz in the magazine where Barry could share all of this wonderful information he was digging up. A lot of magazines run quizzes. We didn't want ours to be a simple rip-off of FAA questions and neither did Barry. "If the answer isn't surprising, educational, or entertaining, I don't want it," Barry insisted. And that has become the mantra surrounding "Test Pilot," *AOPA Pilot*'s monthly quiz that debuted in the March 1994 issue. Every month ever since, our readers have been exposed to a dozen or more questions designed to make them think, or to at least spark a lively debate around the hangar.

What a way to draw out the nitpickers of the world! Readers just love to catch Barry in a mistake. There have been a few, but more often than not the most pious letter writer finds himself humbled when Barry responds with undisputable facts that debunk some of aviation's longstanding myths.

Some questions are purely entertainment: Which lightplane manufacturer designed an automobile immediately after the end of World War II in an effort to offset canceled military contracts and a dim outlook for general aviation? Beech, Cessna, Piper, none of the above. (Hint for eliminating one: The car was called The Plainsman. Beech and Cessna are both located in Wichita.) Other questions will put you in closer touch with your airplane the next time you go flying: An airplane in a normal, climbing turn is… yawing; yawing and pitching; yawing, pitching, and rolling; yawing and rolling? (Hint: The longer answer is usually the right one.)

Over the years, our readers have benefited from and enjoyed "Test Pilot." Now you can too with the questions and answers conveniently grouped into this one edition. So go ahead, turn the page and get started. Keep score if you want, but only you need know how little or much you really know about aviation.

Thomas B. Haines
Editor in Chief, AOPA Pilot magazine

Introduction

The roots of this book took hold in 1993 during the AOPA Expo in Orlando while I was having lunch with the then-editor in chief of *AOPA Pilot*, Mark Twombly. I repeated for the umpteenth time my request to write a quiz for the magazine.

"Pilots enjoy quizzes," I argued.

"That's the problem, Barry. Other magazines already have them, and I don't want *AOPA Pilot* to look like all the others. But I'll think about it." That was Twombly's polite way of changing the subject.

The next day I sat on the panel of "Meet the Writers," a session that gave AOPA members an opportunity to chat with those who write and produce the magazine. When Twombly opened the meeting to questions, a gentleman in the audience wanted to know why our magazine did not have a quiz. I could have kissed the guy. Twombly returned my smug smile with a suspicious leer and later asked how much I had paid the gentlemen.

Twombly later caved, "Okay. You can do the quiz, but I don't want it to be like all the others. It has to be different." I assured him that what I had in mind would indeed be unique.

So was born "Test Pilot," a monthly quiz that debuted in the March, 1994 edition of *AOPA Pilot* and thankfully has gained popularity ever since.

"Test Pilot" has not been popular with everyone, however. Some complain that the quiz is too difficult. But that is precisely its purpose. If the questions could be answered easily, they would serve little purpose other than to assuage the egos of those who participate. Instead, each question and answer is designed to entertain and/or to educate. (I would be disappointed to learn that someone actually scored 100 percent on any of these monthly quizzes.)

Other readers truly enjoy sinking their teeth into "Test Pilot." While attending an EAA AirVenture Fly-In Convention at Oshkosh a few years ago, I met Randy Dunham of the Leading Edge Flying Club in Waterloo, Iowa. He told me that each club member independently takes the quiz before each of the club's monthly meetings, but peeking at the answers is not allowed.

At the meeting, members spend an hour or more discussing each question and arguing their positions. After some changes of opinion, the answers in the back of the magazine are consulted. "What counts most," Dunham claimed, "are not the individual scores, but how much we learn."

That is the spirit in which "Test Pilot" was created.

The book you hold is the evolution of the monthly quiz and contains 1,001 of what I consider to be best of what has been published over the years. They are the most informative, entertaining, and challenging of the lot. I am delighted that this material shall now be preserved for future enjoyment.

Believe it or not, the monthly quiz is more difficult to create than the questions are to answer. Substantial research is required. It helps, however, to be an incurable trivia addict.

When I was a young flight instructor, I always enjoyed discovering nuggets of information and discussing them with others. Not everyone was interested, so this did not always go over well. The owner of the flight school where I worked, Paul N. Bell, used to tell others with tongue-in-cheek (I think) that "Barry knows more [stuff] that nobody gives a [darn] about than anyone I know." But people do care. There is no such thing as useless knowledge, especially to pilots with a passion for aviation and its related subjects.

The exchange of such information also provides fascinating fodder for discussion on the flight decks of long-range airline flights. Airline pilots seem to most enjoy geographical challenges. A popular one is: A pilot takes off from the Detroit City Airport in Michigan and heads due south on a long-distance flight. What is the first foreign country over which he will fly (Question 330)?

I confess to having made a few mistakes in "Test Pilot" over the years even though I go to great pains to ensure accuracy. (Sometimes, I have found, there are errors in otherwise reputable sources.) I pay for these mistakes by personally answering each of the letters that pour in. As a result, I have developed great respect for the expertise of *AOPA Pilot*'s readership.

What I enjoy most, however, are letters asking questions I cannot answer. They force me to research the topic, which can be a fascinating learning experience unto itself. A typical and intriguing example is Question 184, which was posed by Frederick G. Pappas, Jr.

The 1,001 questions in this book include 75 submitted by others, and I have listed these contributors in the appendix.

I readily acknowledge that it is becoming increasingly more difficult to develop appropriate questions for my ongoing monthly quiz, and I would be grateful for any suggestions for questions that you might care to submit. (I'd be most grateful if you would also provide the answers.)

This book is not the product of an individual effort for there is no way that I could have developed this work on my own. I have had much help along the way and wish to publicly express my gratitude to them.

- My close friend, Hal Fishman, who volunteers each month to carefully read the draft of my monthly quiz and check it for accuracy and clarity.
- The editorial staff at *AOPA Pilot* who question my questions and help to keep me on the straight and narrow.
- The management and editorial staff at ASA for having the faith in this material to publish it in book form.
- Jennifer Trerise, my editor at ASA, for her talent, dedication, and the hours that might otherwise have belonged to her.

It seems appropriate to end this introduction with a question that titillates those with a curious nature.

While visiting the Greenwich Observatory near London many years ago, I was fascinated with the brass stripe that runs north-south across the fabled courtyard. It represents the Greenwich Meridian, where one can stand astride the line with one foot in each hemisphere.

An elderly, bespectacled caretaker walked up to me and revealed his obsession with the subject of time (perhaps because he appeared to have so little left).

Without introduction or hesitation, he asked, "Young man, when a given date is first born on Earth, can you tell me how long it will last?"

"Sure. That's easy." I said. "Twenty-four hours."

"No, sir. Every day survives for 48 hours. You think about that, sir."

The old man was pleased with my obvious confusion, pulled out an ancient pipe, lit it, and walked slowly away, circles of smoke rising lazily above his head.

But a moment later, I smiled, too. The old man was right. Can you explain why without looking at the answer to Question 367?

Barry Schiff
Los Angeles

CHAPTER 1
HISTORY

Mix 'n' Match the first and last names:

1._______	Hobbs	A.	Daniel
2._______	Venturi	B.	Gaspard
3._______	Bernoulli	C.	Paul
4._______	Jeppesen	D.	(not a person's name)
5._______	Immelmann	E.	Gerardus
6._______	Chandelle	F.	Elrey
7._______	Pitot	G.	Ernst
8._______	Coriolis	H.	John W.
9._______	Mach	I.	Gabriel
10._______	Kollsman	J.	Max
11._______	Fahrenheit	K.	Henri
12._______	Mercator	L.	Giovanni

____ **Q2**

John Magee wrote the famous poem, *High Flight,*

    A. in a notebook while riding on a Lancaster bomber.
    B. on an envelope while flying a Spitfire.
    C. on a napkin at an airport restaurant.
    D. in a German prisoner-of-war camp.

____ **Q3**

True or False? The first powered flight of an aircraft occurred in 1903.

____ **Q4**

Why are aircraft position lights also called navigation lights?

Which lightplane manufacturer designed an automobile immediately after the end of World War II in an effort to offset canceled military contracts and a dim outlook for general aviation?

A. Beech
B. Cessna
C. Piper
D. None of the above.

Which of the following airplanes holds the world's altitude record of 56,046 feet MSL for piston-powered airplanes?

A. Cessna T210
B. Beechcraft B36TC Bonanza
C. Lockheed 1649A Constellation
D. Fabric-covered Caproni biplane

True or False? On April 3, 1844, a prestigious newspaper, the *New York Sun*, carried the astounding news that the Atlantic Ocean had been crossed by air in three days. This story—called the greatest air hoax ever known—was perpetrated by Edgar Allan Poe.

True or False? The *stick* in the term "dead-stick landing" refers to the control stick (or wheel).

Why is it said when a pilot has a fatal accident that he has "bought the farm"?

Why is a life vest also called a Mae West?

The record for endurance flights (longest time in the air without landing) was flown in

    A. a Boeing 747-400.
    B. the Rutan Voyager.
    C. a helium-filled balloon.
    D. a Cessna 172.

Which of the following does not belong?

    A. king
    B. queen
    C. egg
    D. yoke

The first person to fly an airplane across the Atlantic Ocean was

    A. Albert C. Read.
    B. Charles A. Lindbergh.
    C. John W. Alcock.
    D. Charles Nungesser.

True or False? Charles A. Lindbergh was known as "Lucky Lindy." One reason for this is that during his solo flight across the Atlantic to Paris, the net drift was zero.

Mix 'n' Match the following names with the most appropriate clues:

1. Ellen Church __________
2. Jacqueline Cochran __________
3. Amelia Earhart __________
4. Ruth Nichols __________
5. Harriet Quimby __________
6. Helen Ritchey __________

A. First licensed woman pilot in the United States
B. First U.S. stewardess
C. Established numerous aviation records
D. First woman to fly solo across the Atlantic Ocean
E. Headed Women's Air Force Service Pilots (WASP)
F. First woman airline pilot (Central Air Lines, 1934)

True or False? The first nonstop flight across the United States occurred before the first nonstop flight across the Atlantic Ocean.

True or False? The first horseless carriage was powered by kites.

Mix 'n' Match the following names with the most appropriate clues:

  1. Wilbur Wright __________
  2. Louis Blériot __________
  3. Harold Gatty __________
  4. Glenn Curtiss __________
  5. George Cayley __________
  6. Orville Wright __________
  7. Wiley Post __________
  8. Lowell Smith __________
  9. Otto Lilienthal __________
10. Charles Lindbergh __________

  A. father of aerodynamics
  B. glider pioneer
  C. first airplane across the English Channel
  D. first flight around the world
  E. first successful airplane flight
  F. elder brother
  G. navigator with Wiley Post
  H. first solo flight around the world
  I. first pilot's license in the United States
  J. N-X-211

Why are the control sticks of an airplane (as well as certain computer controls) called joysticks?

What is the name of that mythical haven of rest for the departed souls of aviation heroes?

Modern aircraft are equipped with sensitive altimeters. What was a non-sensitive altimeter?

Four aviation classics, *AOPA Pilot* (the magazine), *Fate is the Hunter* (Ernest K. Gann), *The Spirit of St. Louis* (Charles A. Lindbergh), and *Stick and Rudder* (Wolfgang Langewiesche) were first published, respectively, in

   A. 1944, 1953, 1958, and 1961.
   B. 1958, 1961, 1953, and 1944.
   C. 1961, 1958, 1944, and 1953.
   D. 1953, 1944, 1961, and 1958.

Ernest K. Gann, author of *Fate is the Hunter*. (Q22)

In 1912, which country had the most licensed pilots?

   A. Argentina
   B. England
   C. France
   D. United States

True or False? Victor airways are named after Elias T. Victor, who developed the U.S. airway system.

The total cost to design and custom build the *Spirit of St. Louis* was

    A. $5,580.
    B. $10,580.
    C. $20,580.
    D. $40,580.

The world endurance record for a model glider in free flight is

    A. 1 hour 2 minutes.
    B. 2 hours 47 minutes.
    C. 3 hours 39 minutes.
    D. 4 hours 58 minutes.

Mix 'n' Match the first and last names:

1._______	Cessna	A. Walter
2._______	Piper	B. John
3._______	Beech	C. Edward
4._______	Grumman	D. Claude
5._______	Messerschmitt	E. Clyde
6._______	Bellanca	F. Anthony
7._______	Stinson	G. Al
8._______	Fokker	H. Giuseppe
9._______	Stearman	I. William
10._______	Ryan	J. Willy
11._______	Northrup	K. Lloyd
12._______	Mooney	L. Leroy

Why is such a joyous-sounding word as *mayday* used to sound an emergency?

_______________________________________ **Q 29**

True or False? The *Winnie Mae*, which was the aircraft used twice by Wiley Post to fly around the world in 1931 and 1933, was equipped with a groundspeed indicator.

_______________________________________ **Q 30**

Amelia Earhart was the first woman to fly solo across the Atlantic Ocean from west to east, but who was the first woman to do so in the opposite direction?

_______________________________________ **Q 31**

True or False? The word *aileron* is of French origin and refers to the steering feathers of a small bird.

_______________________________________ **Q 32**

What well-known Hollywood actor had his own television series and also was known for having been issued the first flight instructor certificate?

_______________________________________ **Q 33**

Who was the first man and the first woman to officially break the "sound barrier"?

_______________________________________ **Q 34**

True or False? Most of those aboard the last flight of the *Hindenburg* survived.

_______________________________________ **Q 35**

True or False? The word *empennage* is derived from French and refers to the feathers of an arrow.

Which of the following does not belong?
- A. 12-hp engine
- B. no ailerons
- C. 220 feet
- D. 12 seconds

Mix 'n' Match the following names:

1._______	Celsius	A. Anders
2._______	de Havilland	B. Andrei
3._______	Doppler	C. Christian
4._______	Hawker	D. Donald
5._______	Hertz	E. Ed
6._______	Junkers	F. Geoffrey
7._______	Luscombe	G. Harry
8._______	Sopwith	H. Heinrich
9._______	Swearingen	I. Hugo
10._______	Tupolev	J. Thomas

What is a Cooper switch?

True or False? The throttles on British aircraft before World War II worked backward (i.e., pushing them forward reduced power, and vice versa).

In what year did a Goodyear blimp first fly over a sporting event?
- A. 1925
- B. 1930
- C. 1935
- D. 1940

Lindbergh's historic flight from New York to Paris originated at Roosevelt Field on Long Island. Today this parcel of land is

A. still an airport.
B. a housing tract.
C. an industrial complex.
D. a shopping mall.

Why are long flap and landing gear handles (and other similar levers) often referred to as Johnson bars?

Name the three astronauts who were aboard Apollo 11, the first mission to place men on the moon.

True or False? America's first certificated lady pilot, Harriet Quimby, was killed when she fell out of her airplane.

True or False? As recently as 1975, regulations required that a married female (and not a married male) under 21 years of age who applied for a student pilot certificate provide the written consent of her husband irrespective of his age.

What very famous American entertainer was the first person to fly an airplane in Australia?

A. William "Hopalong Cassidy" Boyd
B. Oliver Hardy (of Laurel and Hardy)
C. Tom Mix
D. Erich Weiss

Mix 'n' Match these popular aviation movies with the most appropriate clues.

1.________ *Airport*
2.________ *Airport '77*
3.________ *Battle of Britain*
4.________ *The Bridges at Toko-Ri*
5.________ *Captains of the Clouds*
6.________ *Command Decision*
7.________ *Fate is the Hunter*
8.________ *Flying Tigers*
9.________ *The High and the Mighty*
10.________ *Memphis Belle*
11.________ *The Spirit of St. Louis*
12.________ *Thirty Seconds Over Tokyo*
13.________ *Those Magnificent Men in Their Flying Machines*
14.________ *Twelve O'Clock High*

A. Royal Canadian Air Force
B. Coffee spill
C. Navy Panthers
D. Short-field takeoff
E. Clark Gable
F. Snowbound
G. "never have so few…"
H. Precursor of Reno
I. Warhawk
J. Point of no return
K. Flying Fortress
L. Underwater pressurization
M. Up periscope
N. 25 missions

_____________________________________ Q 48

Mix 'n' Match these famous aviators with the most appropriate clues.

   1. John Alcock __________
   2. Bernt Balchen __________
   3. Floyd Bennett __________
   4. Amelia Earhart __________
   5. Harold Gatty __________
   6. Fred Noonan __________
   7. Jean-Francois Pilâtre de Rozier __________
   8. Albert Read __________
   9. Alberto Santos-Dumont __________
  10. Wilbur Wright __________

   A. Misled Amelia?
   B. First transatlantic nonstop
   C. First 360-degree turn
   D. First man to fly
   E. First across the Atlantic
   F. First flight over South Pole
   G. *Winnie Mae* navigator
   H. First airplane fight in Europe
   I. First flight over North Pole
   J. First trans-Atlantic passenger

_____________________________________ Q 49

Where did the world's worst air disaster (in terms of lives lost) occur and what type(s) of aircraft was (were) involved?

_____________________________________ Q 50

Why did pilots of a previous generation need to learn the mnemonic, "When Undertaking Very Hard Routes, Keep Directions By Good Methods"?

          Questions

The official U.S. national record for the most consecutive inside loops is

    A. 542.
    B. 1,003.
    C. 1,575.
    D. 2,368.

National Aviation Day is observed on

    A. February 4th (Charles Lindbergh's birthday).
    B. April 16th (Wilbur Wright's birthday).
    C. August 19th (Orville Wright's birthday).
    D. December 17th (first-flight anniversary).

Mix 'n' Match these famous aviators with the most appropriate clues.

1.________ Douglas Bader	A.	Curtiss Robin
2.________ Lincoln Beachey	B.	First across the U.S.
3.________ Gregory Boyington	C.	First "blind" landing
4.________ Paul Carnu	D.	First helicopter flight (1907)
5.________ Douglas Corrigan		
6.________ Glenn Curtiss	E.	First seaplane pilot
7.________ James Doolittle	F.	Glider pioneer
8.________ Icarus	G.	High flier
9.________ Charles Kingford-Smith	H.	Legendary stunt pilot
	I.	Legless fighter pilot
10.________ Otto Lilienthal	J.	Lion tamer
11.________ Umberto Nobile	K.	*Little Boy*
12.________ Calbraith Rodgers	L.	Polar dirigible
13.________ Paul Tibbetts, Jr.	M.	Shepherd
14.________ Roscoe Turner	N.	U.S. to Australia

Some researchers contend that Lieutenant Commander Richard E. Byrd and his pilot, Floyd Bennett, did not get to the North Pole on their famous flight of May 9, 1926. If this contention is correct, who was the first to fly over the pole and in what aircraft?

Who was first to fly a heavier-than-air aircraft?

Why did all transatlantic air service terminate in May, 1937?

With what kind of deicing equipment was Douglas "Wrong-Way" Corrigan's Curtiss Robin equipped during his famous misadventure across the Atlantic Ocean in 1938?

What aerobatic maneuver did *Voyager* perform during its historic, nonstop flight around the world?

Assume that Howard Hughes had asked his mechanic to change all the spark plugs in the Hughes-Kaiser HK-1 *Spruce Goose.* How many plugs would the mechanic had to have changed?

How did the Ninety-Nines get its name?

What is the origin of the word *fuselage*?

_______________________________________ **Q 62**

Francis _______ and Amos _______ worked for Sam Colt
building revolvers. They defined the industrial standard for
the inch and are household names in aviation.

_______________________________________ **Q 63**

Waco and Aeronca are famous names in aviation. They are
also acronyms. What do the letters in these names represent?

_______________________________________ **Q 64**

In 1935, the famous humorist, _______, and noted around-
the-world pilot, _______, were killed while departing a bay near
Pt. Barrow, Alaska in a Lockheed floatplane.

_______________________________________ **Q 65**

True or False? The Wright Brothers never flew together in the
same airplane.

_______________________________________ **Q 66**

How did rain make it possible for Louis Blériot to complete
his historic first flight across the English Channel?

_______________________________________ **Q 67**

The *Spirit of St. Louis* was not equipped with
    A. a sensitive altimeter.
    B. fuel quantity indicators.
    C. a vertical speed indicator.
    D. All of the above.

_______________________________________ **Q 68**

How did it come to be that safety belt buckles in aircraft connect at the front and center of a pilot's torso instead of on the
side (as is the case with automobiles)?

Mix 'n' Match the following individuals with the items to which they are most closely associated:

1._______	Glenn Curtiss	A. aileron
2._______	Philip Dalton	B. autogyro
3._______	James Doolittle	C. barometer
4._______	Theodore Fujita	D. balloon
5._______	Leonardo da Vinci	E. chart
6._______	Juan de la Cierva	F. E6-B computer
7._______	Johann Lambert	G. glider
8._______	Otto Lilienthal	H. helicopter
9._______	Nicholaus Otto	I. instrument approach
10._______	Auguste Piccard	J. internal combustion engine
11._______	Wiley Post	K. jet engine
12._______	Igor Sikorsky	L. microburst
13._______	Evangelista Torricelli	M. parachute
14._______	Frank Whittle	N. pressure suit

Why is the pilot's compartment of an airplane referred to as a cockpit?

What pair of famous brothers, who also were machinists, built the first successful man-carrying aircraft?

After witnessing a successful balloon flight in France, this well-known individual was asked, "What good is it?" The now-famous response was, "What good is a newborn baby?" Who was this person?

True or False? The world duration record for a rubber-band-powered model airplane flown indoors is almost an hour.

Q 74

Charles was a wealthy sportsman and engineer who was eventually killed in a Wright Flyer; Henry struggled from humble beginnings to become an engineer. Their first creation went into production in 1907. What are the last name(s) of this famous duo?

Q 75

Why did early pilots refer to their autopilots as "George" (as in "let George do it")?

Q 76

How many of the first 12 FAA administrators can you name?

Q 77

Which of the following does not belong?
  A. amber
  B. green
  C. magenta
  D. red

Q 78

True or False? The world's first scheduled airline flight was operated using a hot-air balloon.

Q 79

An airport that is closed to VFR pilots because of low ceiling or visibility is said to be "socked in." How did this expression originate?

William was born a Ukrainian peasant in 1895 and became a wealthy American industrialist by inventing and manufacturing a "quarter-turn-to-lock" fastener found on almost all airplanes. His last name was

   A. Dzeus.
   B. Dzus.
   C. Zeus.
   D. Zuse.

Q 81

In what year did the first nonstop flight around the world occur? (Try to guess within 5 years.)

Q 82

Who was first to perform an outside loop?

Q 83

Because of his nerve, what pilot was inducted as a life member of the Wisconsin Liar's Club?

Q 84

Who was the first woman to fly solo around the world?

Q 85

Why does the registration number of U.S. aircraft begin with the letter "N" instead of some other letter (such as a "U")?

Q 86

True or False? An Englishman departed Montreal in a glider (not a motorglider) and flew it across the Atlantic Ocean to southern England.

_______________________________________ **Q 87**

Why is it customary for the pilot-in-command of an airplane
to sit on the left instead of on the right?

_______________________________________ **Q 88**

Why is it customary for the pilot-in-command of a helicopter
to sit on the right instead of on the left?

_______________________________________ **Q 89**

His father manufactured player pianos and nickelodeons, and he
used the innards and principles of organ construction to invent a
revolutionary device in 1929 that today is commonly used and continues to evolve. Who was this pioneer pilot and aircraft owner?

_______________________________________ **Q 90**

Why is seizing control of an aircraft by force referred to as a hijack?

_______________________________________ **Q 91**

One of the most incredible flights in the history of aviation took
place on June 23, 1954 in Santa Monica, California. Identify
that flight.

_______________________________________ **Q 92**

Orville Wright was the first man to solo an airplane, but who
was the first woman?

_______________________________________ **Q 93**

Why was Roger's name chosen to mean, "I have received all of
your last transmission," during radiotelephone communications?

_______________________________________ **Q 94**

Until the late 1940s, U.S.-registered aircraft had registration numbers prefixed by NC-, NX-, NR-, and NS-. What did each of these
prefixes indicate?

In 1921, she overcame the disadvantage of gender and race to become the first black woman pilot in the United States (and probably the world). What was her name?

Mix 'n' Match each of the famous people listed below with the type of airplane in which they perished:

1.________ Ron Brown (U.S. Secretary of Commerce)
2.________ Roberto Clemente (baseball player)
3.________ Patsy Cline (country singer)
4.________ John Denver (singer)
5.________ Dag Hammarskjöld (U.S. Secretary General)
6.________ Buddy Holly, Richie Valens, and the "Big Bopper"
7.________ Rocky Marciano (world boxing champion)
8.________ Glenn Miller (band leader and composer)
9.________ Thurman Munson (baseball player)
10.________ Audie Murphy (World War II hero and actor)
11.________ Ricky Nelson (singer)
12.________ Knute Rockne (legendary football coach)
13.________ Will Rogers (humorist)
14.________ Mike Todd (motion picture producer)

A. Aero Commander
   (piston twin)
B. Beechcraft Bonanza
C. Boeing 737 (USAF CT-43)
D. Cessna 172
E. Cessna Citation
F. Douglas DC-3
G. Douglas DC-6B
H. Douglas DC-7CF

I. Fokker F-10 Trimotor
J. Lockheed Lodestar
K. Lockheed Orion-Explorer
   (on floats)
L. Noorduyn Norseman
   (USAAF UC-64)
M. Piper Comanche
N. Rutan Long-EZ

Who were the only two heads of state to become pilots while holding office?

True or False? Fred Noonan was one of two men aboard the first civilian flight around the world in an airplane.

"Balls to the wall" means to exert maximum effort. How did this expression originate?

Why was it decided in the 18th century that the territorial limit of the United States (and other countries) should extend three miles beyond shore?

The earliest application of an exhaust-driven turbocharger in the United States occurred in

    A. 1908.
    B. 1918.
    C. 1928.
    D. 1938.

A prominent feature of the *Spirit of St. Louis* was a mast mounted on top of the fuselage behind the wing. It supported a set of small cups that spun in the relative wind during flight. This device was used to

    A. determine magnetic heading.
    B. generate electrical power.
    C. measure indicated airspeed.
    D. provide wind and drift information.

Who was an elected member of the U.S. House of Representatives?

A. William Boeing
B. Douglas "Wrong-Way" Corrigan
C. Glenn Curtiss
D. Charles A. Lindbergh

True or False? No American president has ever been a U.S.-certificated pilot.

The first nonstop flight across the Atlantic Ocean was flown by Britons John Alcock and Arthur Brown in a Vickers Vimy biplane in 1919. Who was first to fly nonstop across the Pacific Ocean?

Why is a speed of one nautical mile per hour called a knot?

Mix 'n' Match these "famous firsts" with the year in which each event occurred. If unable to match them, simply list the events in chronological order.

1._______ First flight of a powered airplane
2._______ First powered flight of a manned helicopter
3._______ First flight across the English Channel
(in an airplane)
4._______ First flight across the United States
5._______ First nonstop flight across the Atlantic Ocean
6._______ First flight across the United States in less than a day
7._______ First flight around the world
8._______ First flight over the North Pole
9._______ First nonstop flight from New York to Paris
10._______ First takeoff and landing using instruments only
11._______ First solo flight around the world
12._______ First flight of a jet-powered airplane
13._______ First supersonic flight

A. 1903
B. 1907
C. 1909
D. 1911
E. 1919
F. 1922
G. 1924

H. 1926
I. 1927
J. 1929
K. 1933
L. 1939
M. 1947

Why is the structure used to store an airplane called a hangar?

How did the helicopter get its name?

What is the origin of the term *dead reckoning*?

True or False? A reciprocating aircraft engine remains stationary inside its cowling, and the crankshaft (to which the propeller is rigidly attached) turns within the engine. There were popular aircraft engines, however, that operated the other way around. The crankshaft remained fixed in position (rigidly attached to the airframe), and the engine (to which the propeller was rigidly attached) did the turning.

Nobel laureate, Dr. Alexis Carrel was one of the two co-inventors of the artificial (mechanical) heart. The other was famed pilot

A. Douglas "Wrong-Way" Corrigan.
B. James "Jimmy" H. Doolittle.
C. Howard Hughes.
D. Charles A. Lindbergh.

True or False? The Cuban eight was first performed and demonstrated by a Frenchman, and the chandelle was first performed and demonstrated by a Cuban.

Early airplane designs, military fighters, and current utility/aerobatic airplanes have a control stick that makes it easier to maneuver an airplane than does a control wheel. Why, then, do most lightplanes have control wheels?

What was driven by a propeller, could accelerate to and maintain 143 mph, but could not get off the ground?

I was considered the greatest aviator of all time, but today I am the "forgotten father of aerobatics." Orville Wright called impossible what I could do with an airplane even though 17 million Americans witnessed it during my 126-city tour in 1914. Pilots died trying to duplicate my feats. Thomas Edison said that my exhibition flying was a great contribution to science. Who am I?

The earliest airports had landing areas that were shaped in the form of

    A. large circles.
    B. large rectangles.
    C. large squares.
    D. strips or runways shaped as they are today.

For once you have tasted flight,
You will walk the earth with your eyes turned skyward;
For there you have been,
And there you long to return.

This well-known and beautiful piece of prose was written by

    A. Richard Bach.
    B. Leonardo da Vinci.
    C. Antoine de Saint-Exupéry.
    D. Ernest K. Gann.

Which of the following was (were) the only aviation movie(s) to have won an academy award for being the Best Picture of the Year in which it (they) was (were) nominated?

    A. The High and the Mighty
    B. The Spirit of St. Louis
    C. Twelve O'Clock High
    D. Wings

A person can take dance instruction, martial arts instruction, or various other types of personalized instruction. Why, then, is it said that a student pilot takes *dual* instruction?

What 1904 ground vehicle was the direct result of the Wright Brothers' first flight in 1903?

Why are the enclosures for engines located remotely from the fuselage (as on conventional multiengine airplanes) called *nacelles*?

When was a radio first used aboard an aircraft?

What is the oldest continuously operated airport in the United States?

Westinghouse developed the chicken gun (later called a rooster booster) in 1943. What is it and for what is it used?

The gosport was the first intercom system. It consisted of a flexible speaking tube that enabled the instructor in the front cockpit of an open-cockpit airplane to speak to his student in the rear cockpit by yelling mightily into the funnel at the talking end of the tube. Why was it called a gosport?

____ **Q 127**

What is the largest number of people ever to fly together on
one flight?

    A. 984

    B. 1,084

    C. 1,087

    D. 1,187

____ **Q 128**

True or False? Thirteen flights carrying 91 people crossed the
Atlantic Ocean before Charles A. Lindbergh did it in the *Spirit
of St. Louis* in 1927.

____ **Q 129**

Why are many aviation business (such as flight schools) on an
airport referred to as fixed-base operators?

____ **Q 130**

When an air traffic controller asks, "How do you read?" a pilot
might be heard to reply colloquially with "five-by-five" or "five
square." What do these numbers specifically signify?

____ **Q 131**

A pilot declaring an emergency might transmit "Mayday." The
Aeronautical Information Manual recommends that a pilot with
an urgent condition (less serious than an emergency or distress
situation transmit) "Pan Pan." What is the origin of the word *pan*?

____ **Q 132**

Why is a maximum-performance climbing turn combined with
a 180-degree change in direction called a *chandelle*?

____ **Q 133**

True or False? During aviation's days of yore, pilots flying
aircraft without flaps lost excess altitude on final approach by
using a maneuver called a falling leaf. This consisted of a series
of alternate-direction spin entries and recoveries.

Mix 'n' Match the following aviation writers with their respective works.

1.____	Richard Bach	A.	*Fate is the Hunter*
2.____	Robert Buck	B.	*Flight of the Intruder*
3.____	Stephen Coonts	C.	*Flying Know-How*
4.____	Ernest Gann	D.	*High Flight*
5.____	Assen Jordanoff	E.	*Jonathan Livingston Seagull*
6.____	Wolfgang Langewiesche	F.	*Song of the Sky*
7.____	Charles Lindbergh	G.	*Stick and Rudder*
8.____	John Magee	H.	*Through the Overcast*
9.____	Guy Murchie	I.	*We*
10.____	Antoine de Saint-Exupéry	J.	*Wind, Sand, and Stars*

John Alcock and Arthur Whitten Brown were first to fly nonstop across the Atlantic Ocean (Newfoundland to Ireland, June 14 to 15, 1919). Three weeks later, another aircraft repeated the feat and also became the first to make a round trip across the Atlantic (nonstop each way). What was the aircraft and who was the pilot?

What is the origin of the word *aviation*?

Women pilots were derisively called "petticoat pilots" and "sweethearts of the air," which is why I received little recognition for winning the first Women's Air Derby [the Powder Puff Derby] in 1929. But male pilots did notice when I won the prestigious Bendix Air Race in 1936 against an otherwise all-male field. My name is __________

A. Jacqueline Cochran.
B. Amelia Earhart.
C. Harriet Quimby.
D. Louise Thaden.

____ **Q 138**

I was a car dealer when I bought a Blériot monoplane in 1911 and taught myself to fly (crashing several times in the process). I became an airshow pilot and joined two other gentlemen to begin an airplane company in 1925. But I wanted to build monoplanes, and they preferred biplanes, so I formed my own company in 1927. Two of my airplanes were the DC-6 and DC-6B. My name is ________________.

____ **Q 139**

True or False? A woman, Hanna Reitsch, was first to fly a jet-powered airplane.

____ **Q 140**

What famous pilot was the first person to be killed as the result of an in-flight bird strike?

____ **Q 141**

True or False? Louis Blériot made the first flight across the English Channel.

____ **Q 142**

If Germany had won World War II, what would have been the longitude of Greenwich, England?

____ **Q 143**

Who was the first active U.S. president to fly in an airplane, and what was the type of aircraft in which he flew?

____ **Q 144**

When the author of this book made his first solo flight in 1954, the cost of a sectional chart was

    A. 25 cents.
    B. 50 cents.
    C. $1.00.
    D. $1.50.

_______________________________________ **Q 145**

True or False? In 1941, a private pilot was not allowed to fly
more than three miles beyond the departure airport at night
unless his single-engine airplane was equipped with certified
landing flares to illuminate a landing site in case of engine failure.

_______________________________________ **Q 146**

Why is that lifesaving device called a *parachute*?

_______________________________________ **Q 147**

Orville and Wilbur Wright are among history's most famous
brothers, but how many Wright Brothers were there?

    A. two
    B. three
    C. four
    D. five

_______________________________________ **Q 148**

Who originated the name *Powder Puff Derby* when referring to
the Women's Transcontinental Air Derby?

    A. Arthur Godfrey
    B. Paul Harvey
    C. Howard Hughes
    D. Will Rogers

_______________________________________ **Q 149**

For what aeronautical purpose was the tower atop the Empire
State Building originally intended?

_______________________________________ **Q 150**

True or False? A flying automobile has been the dream of many
aircraft designers. Although many have been built and test flown,
none has ever been certified by the FAA or its predecessor, the
Civil Aeronautics Administration (CAA).

_______________________________________ **Q 151**

German airship designers obviously knew that hydrogen is flammable and that helium is not. So why was hydrogen used on the ill-fated *Hindenburg* and other German airships?

_______________________________________ **Q 152**

Why is the cluster of engine controls in a cockpit often called a *control quadrant*?

_______________________________________ **Q 153**

True or False? During 1942, the first full year of America's involvement in World War II, there were more airports in the United States than there are today.

_______________________________________ **Q 154**

A flamboyant and famous race pilot of the 1930s, ______________, frequently took along as "co-pilot" his pet ____________. The pet's name was ____________.

_______________________________________ **Q 155**

Many years ago, some aircraft were equipped with engine altimeters. What were these?

_______________________________________ **Q 156**

Louis Blériot made history on July 25, 1909, when he used his Blériot XI Monoplane to make the first airplane flight across the English Channel. Which of the following is not correct?

    A. The flight took 36 minutes, but aircraft endurance was 3 hours 30 minutes.

    B. Paper, not fabric, was used as the structural covering.

    C. The aircraft utilized ailerons for roll control.

    D. The French government stationed a destroyer at midchannel in case Blériot had to ditch.

    E. The flight ended with a crash landing atop the white cliffs of Dover.

————————————————— **Q 157**

Almost every pilot knows that the *Spirit of St. Louis* was a Ryan monoplane, but what was the model designation?

————————————————— **Q 158**

What is the Goldfish Club?

————————————————— **Q 159**

True or False? The diesel engine, which might have future applications in general aviation aircraft, was invented by Rudolf Diesel.

————————————————— **Q 160**

What current manufacturer of popular automobiles was the world's first manufacturer of production turbojet engines?

————————————————— **Q 161**

Estimate within 4 years the year in which the greatest number of aircraft were manufactured in the United States.

————————————————— **Q 162**

Which of the following does not belong (aeronautically speaking)?
   A. chicken
   B. goat
   C. duck
   D. sheep

1. H. John W. Hobbs
2. L. Giovanni Venturi
3. A. Daniel Bernoulli
4. F. Elrey Jeppesen
5. J. Max Immelmann
6. D. Chandelle (not a name)
7. K. Henri Pitot
8. B. Gaspard Coriolis
9. G. Ernst Mach
10. C. Paul Kollsman
11. I. Gabriel Fahrenheit
12. E. Gerardus Mercator

$\mathbf{A}$2

B. Magee began the poem in a Spitfire and finished it on the ground.

$\mathbf{A}$3

False. The first powered flight occurred in 1852 when Henri Giffard flew a dirigible equipped with a 3-hp steam engine.

$\mathbf{A}$4

Legend has it that old-timers used to tell student pilots that the best way to prevent from getting lost at night was to keep their airplane directly between the position light on each wing-tip. This presumably is why such lights came to be known as navigation lights.

A. The automobile was called the Beech Plainsman and featured an air-cooled aircraft engine. The project was abandoned in 1946 when it became apparent that the development of new aircraft models—including the new Model 35 Bonanza—would require all of Beech's available talent.

The Beech Plainsman automobile. (A5)

D. This official record seems to belong in *Ripley's Believe It Or Not* and—incredible as it might seem—has stood unchallenged ever since 1938.

True. Arriving in New York nearly penniless and with a sick wife, Poe concocted the story and sold it to the Sun to raise needed funds.

Answers

$$\text{————————————————— } A\,8$$

False. The expression originated before World War I and refers
to the propeller, which — when failing to produce thrust — has
no more utility than a stick.

$$\text{————————————————— } A\,9$$

Early pilots often landed on farms and usually were required
to pay for damaged crops. Such a pilot was said to have "bought"
part of the farm. From this, pilots who fail to survive are said
to have bought the whole farm.

$$\text{————————————————— } A\,10$$

When inflated, the vests reminded early pilots of the buxom
Hollywood actress.

$$\text{————————————————— } A\,11$$

D. In 1958 and 1959, Robert Timm of Las Vegas and John Cook
of Los Angeles flew a Cessna 172 for 64 days, 22 hours, and
19 minutes. Refueling was accomplished by handing a fuel
hose to one of the pilots as they flew low and slow over a pick-
up truck. What a way to build flying time; the duo logged
more than 1,558 hours in little more than two months.

$$\text{————————————————— } A\,12$$

C. In the old phonetic alphabet (Able, Baker, Charlie, etc.), King,
Queen, and Yoke represented K, Q, and Y. Egg has not been
used in any phonetic alphabet.

$$\text{————————————————— } A\,13$$

A. In 1919, Read flew an 84-knot Curtiss NC-4 flying boat
from Newfoundland to Lisbon via the Azores.

$$\overline{\hspace{6cm}}\ A14$$

True. Meteorologists in New York said this was the first time that such "unusual weather conditions [had] been recorded by weather experts."

$$\overline{\hspace{6cm}}\ A15$$

1. B	4. C
2. E	5. A
3. D	6. F

$$\overline{\hspace{6cm}}\ A16$$

False. John Alcock and Arthur Brown made the first nonstop crossing of the Atlantic in 1919. Oakley Kelly and John Macready made the first nonstop flight across the United States in 1923.

$$\overline{\hspace{6cm}}\ A17$$

True. In 1827, George Pocock, an English schoolmaster, hitched a carriage to two large kites flown in tandem. By steering the lower kite with two cords, he navigated the carriage for considerable distances.

$$\overline{\hspace{6cm}}\ A18$$

1. F	6. E
2. C	7. H
3. G	8. D
4. I	9. B
5. A	10. J

$$\overline{\hspace{6cm}}\ A19$$

The control stick was developed by a man whose last name was Joyce and was originally called a Joyce stick, which became foreshortened with use to joystick.

Answers

____ A20

Valhalla.

____ A21

A non-sensitive altimeter had only one hand, and it made a
complete revolution only once every 10,000 feet.

____ A22

B. *Stick and Rudder* has been in continuous publication since
   1944. No other aviation book has even come close to such a
   lengthy publishing run.

____ A23

C. France had 966 pilots, and the United States had only 193.
   Argentina might not have had any.

____ A24

False. Victor comes from the phonetic alphabet. The airways
are so named because they are based on VHF (VOR) NAVAIDs.

____ A25

B. Consider that this was in 1927.

____ A26

D. This incredible record was set by M. Milutinovic of
   Yugoslavia in 1960.

1. E. Clyde Cessna
2. I. William Piper
3. A. Walter Beech
4. L. Leroy Grumman
5. J. Willy Messerschmitt
6. H. Giuseppe Bellanca
7. C. Edward Stinson
8. F. Anthony Fokker
9. K. Lloyd Stearman
10. D. Claude Ryan
11. B. John Northrup
12. G. Al Mooney

Mayday has nothing to do with the first day of May. It is the Anglicized spelling and pronunciation of the French, *m'aidez*, which means, *help me*, or the latter part of the phrase, *(venez) m'aider*, which means, *(come) help me*.

True. The Gatty groundspeed indicator was a prismatic device that allowed the pilot to look through a sight and determine the speed at which objects on the ground passed beneath the aircraft.

Pioneer bush pilot Beryl Markham, whose story was told in the popular book, *West with the Night*.

False. In French, *aileron* refers to a "small wing," which—in a sense—is what an aileron is.

Robert Cummings, who starred in the television series, *Love That Bob*.

Chuck Yeager and Jacqueline Cochran.

A34

True. 62 of the 97 people aboard escaped from the blazing airship.

A35

True. This partially explains why the stabilizers and their related control surfaces are colloquially referred to as tail feathers.

A36

C. The first powered flight of the Wright *Flyer* flew a distance of 120 feet, not 220 feet.

A37

1. A. Anders Celsius
2. F. Geoffrey de Havilland
3. C. Christian Doppler
4. G. Harry Hawker
5. H. Heinrich Hertz
6. I. Hugo Junkers
7. D. Donald Luscombe
8. J. Thomas Sopwith
9. E. Ed Swearingen
10. B. Andrei Tupolev

A38

Named after the infamous hijacker, D.B. Cooper, it is an air switch that prevents opening the aft (air) stair door of a Boeing 727 in flight.

A39

False. But it was true of French aircraft, and this caused problems for French pilots flying Royal Air Force Spitfires during the Battle of Britain.

A40

A. The aircraft used was Goodyear's first civilian blimp, the diminutive Pilgrim. (*See* photo at right.)

D. Roosevelt Field Mall is in Garden City, NY.

This term originated with the Johnson bar (a tall brake handle) found on old locomotives and was named after the engineer who developed it, T.H. Johnson.

Neil Armstrong, Edwin "Buzz" Aldrin, and Michael Collins. (How quickly we forget.)

Goodyear's first civilian blimp, *Pilgrim*. (A40)

_______________________________________ A44

True. Neither she nor her passenger were wearing safety belts when they were jostled from their seats by turbulence.

_______________________________________ A45

False. This was true, however, in 1965.

_______________________________________ A46

D. The world-renowned magician and escape artist (also known as Harry Houdini) made the flight on March 18, 1910.

_______________________________________ A47

1. F		8. I	
2. L		9. J	
3. G		10. N	
4. C		11. M	
5. A		12. D	
6. E		13. H	
7. B		14. K	

_______________________________________ A48

1. B		6. A
2. F		7. D
3. I		8. E
4. J		9. H
5. G		10. C

_______________________________________ A49

This tragedy occurred in Tenerife in the Canary Islands on March 27, 1977. 582 people were killed when two Boeing 747s (one taxiing and one taking off) collided on the runway.

Rotating beacons were installed along airways at 10-mile intervals. Each flashed a letter in Morse code (in the sequence of the mnemonic) to help a pilot determine his position along an airway at night. The sequence repeated every 100 miles. A few such beacons are still operating in Montana.

D. This incredible record was established by David Childs in a Bellanca Decathlon in 1986.

B. This also is the birthday of Robert "Boom" Powell, who submitted this question.

1. I	8. G
2. H	9. N
3. M	10. F
4. D	11. L
5. A	12. B
6. E	13. K
7. C	14. J

The "second" flight to the North Pole followed the Byrd flight by three days. Roald Amundsen, Lincoln Ellsworth, and Umberto Nobile flew there in the Italian-built dirigible, *Norge*.

According to *American Heritage of Flight*, Sir George Cayley, considered the Father of Aerial Navigation, "coaxed his reluctant coachman [in 1853] into making a trial run in a glider that lifted him across a shallow valley and set him down in a cloud of dust."

--- A56

This is when the *Hindenburg* erupted in flame at Lakehurst, New Jersey, which abruptly ended the use of hydrogen-filled airships. Boeing 314 flying boats enabled the resumption of service in 1939.

--- A57

The airplane was equipped with an 8-foot-long stick. Corrigan would stick it out the window to scrape ice off the wings as it accumulated during flight.

--- A58

An outside loop of global proportion.

--- A59

The *Spruce Goose* had eight Pratt & Whitney R-4360 Wasp Major engines. Each had 28 cylinders and required 56 spark plugs. The 3,000-hp engines, therefore, had a total of 448 spark plugs.

--- A60

The organization of international women pilots was named for the 99 charter members. Amelia Earhart was elected as the first president.

--- A61

Fuselage comes from the French word, *fuselé*, which means *spindle-shaped*.

--- A62

Pratt (and) Whitney

--- A63

Weaver Aircraft COmpany and AERONautical Corporation of America.

Will Rogers, Wiley Post.

False. The brothers made it a firm rule never to fly together to ensure that one of them would be available to continue their work. On May 25, 1910, however, Orville took Wilbur for their only flight together, a tacit concession that most of their work had been completed.

Blériot's overheating, 25-hp Anzani engine probably would have failed had it not been for the cooling effect of a rain shower.

D. Charles Lindbergh kept track of the fuel remaining in each fuel tank by calculating fuel flow vs. time. The altimeter was non-sensitive (a single hand made one counterclockwise revolution every 20,000 feet).

In case of an accident resulting in an injured or broken arm, the injured person can most easily reach a centrally located buckle irrespective of which arm is injured.

1. A	8. G
2. F	9. J
3. I	10. D
4. L	11. N
5. M	12. H
6. B	13. C
7. E	14. K

$$A70$$

A cockpit was the pit in which fighting cocks battled, and came to mean any small place where many battles were waged, as in the adage, "Belgium is the cockpit of Europe." By analogy, World War I pilots in Europe referred to their cramped quarters as cockpits, which came to mean any place on a plane or boat used by the steerman.

$$A71$$

The Montgolfier Brothers, Jacques and Joseph, built and launched a man-carrying balloon in 1783.

$$A72$$

Benjamin Franklin, who was the U.S. diplomatic representative to France at the time.

$$A73$$

True. This remarkable record of 58 minutes 8 seconds was set in 1996 by Stephen Brown and required 2,390 windings of the propeller.

$$A74$$

Rolls and Royce. Their first product was the "Silver Ghost" automobile.

$$A75$$

Early autopilots were gyroscopic systems (or G systems, for short). In the old phonetic alphabet, "George" (instead of "Golf") represented the letter, "G."

The first twelve were

Elwood Quesada	J. Lynn Helms
Najeeb Halaby	Donald Engen
William McKee	T. Allen McArtor
John Shaffer	James Busey IV
Alexander Butterfield	Langhorne Bond
John Lucas	David Hinson

C. Amber, green, red, (and blue) were the designations for the low/medium-frequency airways that preceded Victor (VHF) airways.

False. The St. Petersburg-Tampa Airboat Line used a Benoist flying boat (January 1, 1914) that carried one passenger on a 23-minute flight.

"Socked in" was originally "sock in" and was first used in early French aviation. During inclement weather, the wind sock was dismasted and taken indoors. Also, the colloquial French meaning of "sock" was "close in" or "conceal."

B. The Dzus fastener is the most effective way to join non-structural panels (such as engine cowlings) so as to allow quick assembly and disassembly. (Zeus was a Greek god.)

$$\overline{\hspace{7cm}}\,A81$$

The first nonstop flight around the world was completed on March 2, 1949 by a Boeing B-50A. The *Lucky Lady II* was refueled in flight four times by KB-29 tankers, landed at Carswell AFB, and flew 23,453 SM in 94 hours 1 minute.

$$\overline{\hspace{7cm}}\,A82$$

A little-known Frenchman, Adolphe Pégoud performed an outside loop in a Blériot monoplane on September 1, 1913. The first American was James H. "Jimmy" Doolittle, who used a Curtiss fighter in the same month that Lindbergh flew to Paris (May, 1927).

$$\overline{\hspace{7cm}}\,A83$$

Douglas "Wrong-Way" Corrigan took off from New York in his dilapidated Curtiss Robin in 1938, landed in Ireland instead of California, and claimed that his "compass must have been wrong."

$$\overline{\hspace{7cm}}\,A84$$

In 1964, Geraldine "Jerrie" Mock, a housewife from Columbus, Ohio, used a Cessna 180 to complete the 23,103-SM circumnavigation in less than 30 days.

$$\overline{\hspace{7cm}}\,A85$$

When an international convention met in Paris, France, in 1919 to decide upon matters of national registration, the first-ever flight across the Atlantic had just been completed by the American NC-4 flying boats. The worldwide popularity of these "N-boats" led to the selection of the letter "N" for U.S.-registered aircraft.

$$\overline{\hspace{7cm}}\,A86$$

True. The Waco-designed glider was towed across the Atlantic by a Douglas C-47 and piloted by Squadron Leader R. G. Seys of the Royal Air Force in June 1943 for delivery to Russia.

_______________________________________ **A**87

The custom seems to have evolved from a maritime rule of the road. It states that vessels approaching each other head-on must pass port to port (left side to left side). Sitting on the left afforded the best view of such a passing vessel.

_______________________________________ **A**88

Early helicopters had a single collective/throttle control between the two pilots. Each pilot was provided with a control stick (cyclic). Because it was ergonomically desirable to use the stick with the right hand, this required the pilot-in-command to sit on the right (and use the collective with his left hand).

_______________________________________ **A**89

Ed Link invented the Link trainer, which evolved into today's sophisticated simulators.

_______________________________________ **A**90

The most widely accepted origin of the word hijack stems from "High, Jack," a command given to drivers of trucks loaded with illicit liquor in the 1920s to raise their arms during highway robberies. It is, of course, unwise to say "Hi" to a pilot named Jack.

_______________________________________ **A**91

The author of this book made his first solo flight (in an Aeronca 7AC Champion) without crashing.

_______________________________________ **A**92

According to an article by Richard Bauman, Blanche Stuart Scott, "The Flying Tomboy," soloed a Curtiss airplane on September 2, 1910 and became a professional stunt pilot only two months later.

_______________________________________ A93

The letter, R, was used in early Morse code communications to mean, "I have Received your message." It was natural to use Roger in radiotelephony because it represented the letter R in the first phonetic alphabet (Able, Baker, Charlie, etc.).

_______________________________________ A94

NC- represented civil aircraft. NX- represented experimental aircraft (such as NX-211, the *Spirit of St. Louis*). NR- represented restricted aircraft (such as crop-dusters and racers). NS- represented state-owned aircraft (those belonging to Federal and state governments).

_______________________________________ A95

Bessie Coleman (a.k.a. Brave Bessie) could not find a school in the U.S. that would teach her to fly. She found more liberal attitudes in France, learned to fly there, and returned home to become a successful exhibition pilot.

_______________________________________ A96

1. C. 1996		8. L. 1944	
2. H. 1972		9. E. 1979	
3. M. 1963		10. A. 1971	
4. N. 1997		11. F. 1985	
5. G. 1961		12. I. 1931	
6. B. 1959		13. K. 1935	
7. D. 1969		14. J. 1958	

_______________________________________ A97

King Hussein bin Talal of Jordan and King Michael of Romania.

_______________________________________ A98

False. Fred Noonan was Amelia Earhart's navigator. Harold Gatty was Wiley Post's navigator and, therefore, was one of the first two civilians to fly around the world.

A99

The knobs atop the engine-control levers of several older, multi-engine airplanes (such as the Douglas DC-3) consisted of a cluster of small, marked balls that were pushed fully forward (to the firewall) to produce maximum power. This is similar to the automobile-related expression, "pedal to the metal."

A100

This was the maximum range of a shore-based cannon.

A101

B. This is when General Electric attached a turbocharger to a Liberty engine, which produced more horsepower on the ground at the top of Pike's Peak (14,110 feet MSL) than at sea level and ultimately led to a 1920 altitude record of 33,000 feet.

A102

A. The device was an earth-inductor compass. It was essentially a wind-driven generator that was used to determine the angle between the longitudinal axis of the airplane and the Earth's lines of magnetic force (flux).

A103

D. Charles A. Lindbergh, the father of famed pilot, Charles A. Lindbergh, was a Republican congressman (Minnesota) from 1907 to 1917. (Lindbergh's kidnapped baby was Charles A. Lindbergh, Jr.)

A104

False. President Dwight D. "Ike" Eisenhower had a pilot certificate.

____ A105

Clyde Pangborn and Hugh Herndon, Jr. flew their Bellanca Skyrocket, *Miss Veedol*, from Japan to the United States in 1931 and covered the 4,465 statute miles in 41 hours and 13 minutes.

____ A106

Mariners of yore tied equally spaced knots in a line (rope), threw one end of it overboard, and paid out the line, which floated on the surface. The time required for a given number of knots to float past a given part of the ship was used to calculate the ship's speed. For example, if the knots in the line were 47 $1/4$ feet apart, and it took 28 seconds between the passage of each of these knots, the speed would be 1 nautical mile per hour, which was called a knot.

____ A107

Anyone knowing only a few of these dates might recognize that the "famous firsts" are presented in chronological order.
1.________ Orville Wright (1903)
2.________ Paul Cornu (1907)
3.________ Louis Blériot (1909)
4.________ Cal Rodgers (1911)
5.________ John Alcock and Arthur Brown (1919)
6.________ James Doolittle (1922)
7.________ J. Macready and O. Kelly (1924)
8.________ Richard Byrd and Floyd Bennett (1926)
9.________ Charles Lindbergh (1927)
10.________ James Doolittle (1929)
11.________ Wiley Post (1933)
12.________ Heinkel He-178 (1939)
13.________ Charles Yeager (1947)

____ A108

The word, *hangar*, comes from the French word, *hangar*, which means outhouse or shed.

____ A109

Helicopter comes from the Greek words, *helikos*, which means *spiral*, and *pteron*, which means *wing* (a spiraling wing).

____ A110

There are two prevalent and contradictory schools of thought. The most popular claims that it is a nautical term meaning [navigational] *reckoning relative to an object dead in the water*. Another claims that the expression was originally *deduced reckoning*, which became foreshortened to *d'ed* or *dead reckoning*.

____ A111

True. The most famous were the Le Rhone, Gnome, and Bentley rotary-type radial engines, which were used to power many World War I airplanes. The torque and gyroscopic effects of the large rotating masses of these engines made aircraft control difficult.

____ A112

D. In 1935, Lindbergh and Carrel developed an external pump that could be used to sustain blood circulation during heart surgery. Lindbergh also developed a centrifuge to separate blood plasma without damaging it.

____ A113

False. The first Cuban eight is credited to Len Povey, a colorful American barnstormer of the 1930s who headed the Cuban Air Force and was Cuban President Fulgencio Batista's personal pilot. The first chandelle is credited to French pilot, Maurice Chevillard, a maneuver he originally called a looping helicoidal.

$$A114$$

Control wheels originated in lightplanes as a marketing effort by airframe manufacturers to attract automobile drivers to general aviation. They were more familiar, less intimidating, and more like airliners. Ironically, some modern airliners now incorporate sidestick controllers.

$$A115$$

In 1931, Germany produced a train that was powered by a 500-hp "oil" engine, weighed 18 tons, carried 40 passengers, and was the fastest speed ever achieved on rail.

$$A116$$

Shame on you. I am Lincoln Beachey, and I was first in America to perform a loop and the first anywhere to perform barrel rolls, tail slides, and a host of other aerobatic maneuvers. I was first to fly upside-down and first to achieve terminal velocity during vertical flight. I also am credited with developing stall-recovery techniques. (No one ever accused me of being humble.)

$$A117$$

A. Circular landing areas ensured that a pilot could always land into the wind. This was critical when flying early aircraft because they had limited crosswind capability. The windsock typically was at the center of the circle. Airships, which have virtually no ability to land crosswind, still use circular landing areas.

$$A118$$

B. This Italian master (1452–1519) of the Renaissance Era was writing about the flight of birds.

D. This 1927 World War I classic and Oscar winner starred
   Richard Arlen, Clara Bow, and Buddy Rogers.

A119

Early pilots were given instruction in airplanes with a single set of
controls. When airplanes became available with dual controls,
instruction in such airplanes became referred to as dual instruction.

A120

The Macduff Aeropinion was a propeller-driven automobile
that generated prodigious clouds of dust and pebbles on the
dirt streets. It was dangerous because of its exposed propeller
and went out of production in 1905.

A121

Nacelle is French for "small boat" and refers aeronautically to
any separate and streamlined enclosure on an airplane used to
shelter something (usually an engine).

A122

On September 9, 1910, a British Bristol Boxcar piloted by Robert
Loraine flew near Stonehenge, England and communicated with
the ground using a Marconi radio.

A123

The curators at the museums at both Pearson Field (VUO) in
Vancouver, Washington, and College Park Airport (CGS) in
Maryland lay claim to this title. One of them is apparently correct.
Take your pick.

A124

A125

This compressed-air cannon is used to fire deceased chicken carcasses at the windshields and leading edges of high-speed aircraft to test their ability to endure bird strikes.

A126

This primitive intercom got its name from the British Military Flying School at Gosport, England, which is where the device was first used. (Gosport was used later as a major embarkation point for the Normandy Invasion in 1944.)

A127

B. and C. In 1991, a Boeing 747 operated by El Al Israel airlines took off during an evacuation from Ethiopia with 1084 people on board. There were three births during the flight, which landed in Tel Aviv with 1,087 people on board.

A128

True. Lindbergh was the first to fly nonstop between two major cities (New York to Paris), and he did it solo.

A129

In the early days of aviation, some transient barnstormers collected money in advance for promised flights and then disappeared in the night without satisfying their obligations. This led to the expression, "fly-by-night operations." The expression *fixed-based operator* was the result of honest pilots remaining on the field for extended periods.

A130

High readability and signal strength. A response of "one by three," for example, would signify, "poor readability and moderate signal strength." This aeronautically unofficial phraseology originated during the early days of radio communications. Today, "five by five" is essentially synonymous with "loud and clear."

_______________________________________ $A131$

Pan comes from the French word, *panne*, which means breakdown. Aeronautically, therefore, *pan* signifies a breakdown in safety.

_______________________________________ $A132$

Chandelle is taken from the French expression, *monter en chandelle*, which means "climb on (or around) a candle," and is similar to the American idiom, "turn on a dime."

_______________________________________ $A133$

True. The maneuver is executed by holding the aircraft in a stall and entering and recovering from a series of incipient spins in alternating directions. When viewed from ahead, the aircraft appears to be falling like a leaf.

_______________________________________ $A134$

1. E	6. G
2. C	7. I
3. B	8. D
4. A	9. F
5. H	10. J

_______________________________________ $A135$

Major G. H. Scott and a crew of 30 made the round-robin flight in a 65-foot-long British R.34 dirigible. The westbound leg from Scotland to Mineola, Long Island, took 108 hours, and the return flight to Norfolk, England, took 75 hours.

_______________________________________ $A136$

Aviation is a French word derived from the Latin, avis, which means bird.

D. Quimby was the first woman pilot licensed in the United States and the first woman to fly across the English Channel (1912). Cochran won the Bendix Air Race in 1938, headed the U.S. Women Airforce Service Pilots (WASPs) during World War II, and was the first woman to fly faster than the speed of sound (1953).

A138

Clyde Cessna (and not Donald Douglas) had two partners in the Travel Air Manufacturing Company, Walter Beech and Lloyd Stearman. The DC-6 (D for design, C for Cessna, and 6 for sixth design) was a four-place, high-wing, closed-cabin monoplane that debuted in 1929.

Clyde Cessna and his *Comet.*   (A138)

A139

False. The first turbine-powered airplane was the Heinkel He-178, which was first flown by Erich Warsitz on August 27, 1939. Although not the first, Reitsch did fly the world's first rocket-powered airplane, the Messerschmitt 163. The petite aviatrix also flight-tested a mammoth glider capable of carrying 200 soldiers and a tank.

_______________________________________ A140

Cal Rodgers, first to fly across the United States, was killed in 1912 after a seagull became jammed in the flight controls of his Wright biplane.

_______________________________________ A141

False. On January 7, 1785, Jean-Pierre Blanchard and John Jeffries took two and one-half hours to cross the channel in a balloon.

_______________________________________ A142

13 degrees 15 minutes west. Adolf Hitler had planned to move the world's Prime Meridian (zero degrees longitude) to Berlin, Germany. He even had maps and charts printed to reflect that change.

_______________________________________ A143

President Franklin D. Roosevelt (FDR) flew to Casablanca during World War II in a Boeing 314 seaplane (a "Pan Am Clipper").

_______________________________________ A144

A. Expired charts sold for 10 cents each; an Aeronca Champion rented for $7 per hour (including fuel); the instructor charged $4 per hour; and 80-octane avgas sold for 35 cents per gallon.

_______________________________________ A145

False. Such flares were required if the aircraft were being operated for hire and obviously were effective only if deployed when the aircraft was directly over the intended landing site. They also drifted with the wind during descent and were known to start ground fires, which created more illumination than the government had intended.

_______________________________________ A146

Literally translated from French, _parachute_ means _to guard against a fall._

C. Orville (born 1871) and Wilbur (1867) had two older brothers, Reuchlin (1861) and Lorin (1863)—and a younger sister, Katharine (1874). (There also were twins, who died in infancy.)

D. The famous humorist and Wiley Post were killed in a seaplane accident near Point Barrow, Alaska, in 1935.

The original tower (now at the base of the TV tower) was designed as a mooring mast for transatlantic airships. Gusty winds at 1,350 feet ASL (above street level) resulted in several unsuccessful mooring attempts, and the idea was abandoned.

False. An early certified flying automobile was the Airphibian (1950), which was conceived and designed by Robert Edison Fulton Jr. It took less than five minutes to convert the airplane into an automobile and vice-versa. Another was the Aerocar, which was certified six years later. (See figures below and to right.)

Robert Fulton's Airphibian. (A150)

_______________________________________ A151

The United States had a monopoly on the world production of helium and would not sell this inert gas to the German government.

_______________________________________ A152

On early aircraft (especially multiengine aircraft), engine controls were moved through a 90-degree arc or quadrant. With the obvious exception of push-pull controls, modern engine controls typically move through a smaller arc (less than a quadrant).

_______________________________________ A153

False. It wasn't even close. There were only 1,084 "landing fields" in the 48 states in 1942. As of January 1, 1998, there were 18,345 landing facilities (including heliports, STOLports, and seaplane bases), of which 13,192 were airports.

The Mizar crashed on its first flight at Van Nuys Airport in 1973. (A150)

$$\overline{\qquad\qquad\qquad\qquad\qquad\qquad}\ A154$$

Roscoe Turner's pet *lion* was named *Gilmore*. (Turner's sponsor at that time was the Gilmore Oil Company, the logo of which was a lion.)

$$\overline{\qquad\qquad\qquad\qquad\qquad\qquad}\ A155$$

These were manifold pressure gauges that indicated the altitude corresponding to the pressure produced in the intake manifold of a supercharged engine. For example, a manifold pressure of 28 inches Hg was indicated by an engine altitude of 2,000 feet.

$$\overline{\qquad\qquad\qquad\qquad\qquad\qquad}\ A156$$

C. Like the Wright *Flyer*, wing warping was used on the Blériot XI Monoplane for roll control.

$$\overline{\qquad\qquad\qquad\qquad\qquad\qquad}\ A157$$

N-X-211 was officially designated as a Ryan NYP. The letters stood for *New York to Paris*.

$$\overline{\qquad\qquad\qquad\qquad\qquad\qquad}\ A158$$

It is (or was) an organization of flight crews whose lives have been saved after ditching by the use of emergency dinghies or rafts.

$$\overline{\qquad\qquad\qquad\qquad\qquad\qquad}\ A159$$

True. The French-born German engineer (1858-1913) developed the "pressure-ignited heat engine" between 1885 and 1898.

$$\overline{\qquad\qquad\qquad\qquad\qquad\qquad}\ A160$$

BMW (Bavarian Motor Works) also was a major manufacturer of German reciprocating engines during both world wars.

There were 96,369 aircraft built in 1944, the last full year
World War II.

B. The first successful aerial voyage with a live payload occurred
in France on September 19, 1783. The Montgolfier Brothers
placed a cock, a duck, and a sheep aboard their aerostat (a hot-
air balloon) to determine if the upper air could sustain life,
and wished them bon voyage. The animals returned safely after
an 8-minute flight. An alternate answer is C, because of the
four animals, only the duck can fly.

CHAPTER 2
PILOTING

True or False? The buffeting felt by a pilot during stall entry in most lightplanes is the result of burbling air from the wing striking the tail.

Q 164

Two identical airplanes are climbing at the same density altitude. One, however, is climbing within a low-pressure area and the other within a high-pressure area. Which airplane has the best climb performance?

Q 165

A pilot is flying a perfect circle over the ground while maneuvering counterclockwise around a pylon. The wind is from the north. On which side of the circle is the steepest bank angle required?

A. the north side
B. the east side
C. the south side
D. the west side

Author performing an around-pylon over Mont-Saint-Michel in Northern France. (Q165)

If an airplane flies into an updraft, it most likely will

    A. pitch up.
    B. pitch down.
    C. pitch up or down.
    D. not change pitch.

A pilot notes that the static sources of his unpressurized airplane have become clogged by structural icing while climbing at 500 fpm. He restores operation of his pitot-static instruments by breaking the glass of the vertical-speed indicator. After doing so, he notes that the VSI indicates

    A. a 500-fpm sink rate.
    B. no vertical speed.
    C. a 500-fpm climb.
    D. a 1,000-fpm climb.

An airplane in a normal, climbing turn is

    A. yawing.
    B. yawing and pitching.
    C. yawing, pitching, and rolling.
    D. yawing and rolling.

A pilot is executing a normal turn at a constant altitude. He then steepens the bank angle and increases airspeed. This causes turn rate to

    A. increase.
    B. decrease.
    C. remain the same.
    D. Cannot be determined from the information given.

An airplane with a normal glide speed of 80 knots glides 10,000 feet forward for every 1,000 feet of altitude lost (a glide ratio of 10:1). If the pilot disposes of 1,000 pounds of payload while gliding at the same airspeed, the glide ratio would

A. increase.
B. decrease.
C. remain the same.
D. cannot be determined.

A sadistic instructor places Scotch tape over the static ports of the aircraft to see if his student will notice the defect during the pre-flight inspection. The sabotage, however, goes undetected, and the instructor forgets about the tape. Which statement describes the behavior of the indicated airspeed during the subsequent takeoff roll?

A. IAS is unaffected.
B. IAS is greater than it should be.
C. IAS is less than it should be.
D. IAS remains at zero.

An airplane is parked on a ramp at 5,000 feet MSL when the ambient temperature is 100°F. The altimeter has been properly set to the local altimeter setting and indicates 5,000 feet. That evening, the temperature plummets to 30°F. Unless reset, the altimeter will indicate

A. 5,200 feet MSL.
B. 5,000 feet MSL.
C. 4,800 feet MSL.
D. 4,600 feet MSL.

A normally aspirated, single-engine airplane has a maximum climb rate of 1,500 fpm at sea level and 1,000 fpm at 5,000 MSL. It has a service ceiling of

A. 14,000 feet MSL.
B. 15,000 feet MSL.
C. 16,000 feet MSL.
D. 18,000 feet MSL.

When a pilot flies into the region of reversed command,

A. he will encounter aileron reversal.
B. he will need more power to fly more slowly.
C. he will detect a high-frequency buffet.
D. the aircraft will stall.

How does an increase in density altitude affect best angle-of-climb airspeed (V_X) and best rate-of-climb airspeed (V_Y) in a lightplane with a normally aspirated piston engine?

A. Both speeds increase.
B. Both speeds decrease.
C. V_Y increases and V_X decreases.
D. V_X increases and V_Y decreases.

A pilot maintaining a constant altitude jabs at the right rudder pedal and induces a right skid. This temporarily causes the vertical-speed indicator to show a descent, the altimeter to show an altitude loss, and the indicated airspeed to decrease. These instrument errors are caused by

A. a blocked left static port.
B. a blocked right static port.
C. both static ports are blocked.
D. an open alternate static source.

When taking off in a tricycle-gear airplane, which of the following causes the left-turning tendency prior to rotation?

   A. the propeller's spiraling slipstream
   B. gyroscopic precession
   C. p-factor
   D. torque

True or False? It is possible to fly any light airplane such that it cannot be made to stall at any airspeed.

During cruise flight at a constant altitude in a Cessna 172, a passenger climbs from the front seat to the rear seat and spends the rest of the flight there. This causes

   A. airspeed to increase and stall speed to decrease.
   B. airspeed and stall speed to increase.
   C. airspeed to decrease and stall speed to increase.
   D. airspeed and stall speed to decrease.

An aerobatic pilot decides to be cute and executes a visual approach to Runway 36 while inverted. During the approach, he notes that the VASI lights indicate red and that the aircraft has drifted east of the runway centerline. To correct for these discrepancies, the pilot should

   A. pull the stick aft and move it to his left.
   B. pull the stick aft and move it to his right.
   C. push the stick forward and move it to his left.
   D. push the stick forward and move it to his right.

A pilot enters and maintains a prolonged 45-degree banked turn while also maintaining a constant airspeed and altitude in a typical light airplane. After some period of time, the conventional air-driven attitude indicator will

A. gradually indicate a steeper-than-actual bank angle.

B. continue to indicate a 45-degree bank angle.

C. gradually indicate a shallower-than-actual bank angle.

D. gradually precess in either direction.

A pilot takes off from an airport at a high density altitude in a light airplane with a turbocharged engine that develops approximately the same horsepower that it does at sea level. Nevertheless, takeoff distance is greater than at sea level primarily because

A. the aircraft must accelerate to a greater indicated airspeed.

B. the aircraft must accelerate to a greater groundspeed.

C. of the change in stall speed.

D. of the increased induction temperature due to turbocharger compression.

An aerobatic pilot enters a knife-edge turn to the left. While maintaining this 90-degree bank angle, his conventional turn indicator will show

A. a large turn rate to the left.

B. a large turn rate to the right.

C. either a left or right turn.

D. that the airplane is not turning.

Pitts S-2A
aerobatic biplane.
(Q183)

Before takeoff, a pilot lights a short candle and sets it in a holder on the center of his glareshield. While accelerating for takeoff (and with all the vents closed to prevent air circulation in the cockpit), the pilot notes that the flame

A. leans forward.
B. leans rearward.
C. leans right.
D. leans left.

The same candle (as in the previous question) remains on the glareshield while the pilot performs a positive-G loop at a constant airspeed. While at the top of the loop, the tip of the flame

A. points toward the floor of the airplane.
B. points toward the ceiling of the airplane.
C. points toward the ceiling and leans left.
D. points toward the floor and leans left.

To maximize gliding distance while gliding with a tailwind, a pilot should use

A. the glide speed published in the pilot's operating handbook.
B. an airspeed that is greater than the published glide speed.
C. an airspeed that is less than the published glide speed.

A pilot opts to fly a typical lightplane without the propeller spinner. The most adverse result of this is

A. engine overheating.
B. excessive drag.
C. an adverse effect on weight and balance.
D. hazardous aerodynamic buffeting.

The most important reason not to use carburetor heat while taxiing is that this can

    A. damage the engine.
    B. cause excessive oil temperature.
    C. cause an excessively lean mixture.
    D. cause an excessively rich mixture.

A pilot is taxiing along a narrow taxiway at an uncontrolled airport on a cloudless day when he notices another 172 approaching from the opposite direction. Without stopping, how can both pilots be certain that their wingtips will not touch as they pass one another?

True or False? A conventional, tricycle-gear airplane has a normal 1-G stall speed of 60 knots in the takeoff configuration. When the indicated airspeed of this airplane passes through 40 knots during a normal takeoff roll, the wings are stalled.

True or False? A typical carbureted engine fails during cruise flight because of fuel starvation. Manifold pressure remains unchanged but can be reduced by closing the throttle.

Most light airplanes have a maximum-allowable limit load factor of +3.8 Gs. A pilot will exceed this limit by exceeding a bank angle of _______ degrees during a steep turn at constant altitude.

A pilot applies carburetor heat during his preflight runup, and engine rpm increases. What does this indicate (assuming that there is no ice in the carburetor)?

A pilot begins a normal glide at 10,000 feet MSL and maintains the best glide speed throughout the subsequent descent. As the airplane loses altitude, sink rate

A. increases.
B. decreases.
C. remains the same.
D. cannot be determined.

True or False? During a stabilized gliding turn with a 60-degree bank angle, the load factor (G load) is less than when in a similarly banked turn while maintaining altitude at a higher airspeed.

A flat spin is commonly defined as a spin during which the angle of attack is greater than

A. 15 degrees.
B. 30 degrees.
C. 60 degrees.
D. 75 degrees.

What is a spin called when performed in approximately horizontal flight?

During a power-off approach to landing (assuming that the downwind, base and final-approach legs are of equal length in each of the following cases), the most altitude will be lost when there is a

A. direct headwind on the downwind leg.
B. direct tailwind on the downwind leg.
C. direct headwind on the base leg.
D. direct tailwind on the base leg.
E. 45-degree headwind on final approach.

____ **Q 199**

What technique used by ballet dancers can be used by a pilot to avoid becoming dizzy during a spin?

____ **Q 200**

True or False? The first indications to a pilot that his airplane is entering a low-altitude microburst is that the nose pitches up, and there is an increase in lift.

____ **Q 201**

True or False? A pilot is flying an airplane equipped with a conventional and controllable elevator trim tab. The nosewheel tire blew during takeoff, and during the subsequent landing, the pilot wants the nosewheel to touch down at the lowest possible speed. Therefore, he should gradually apply full nose-down trim while holding the control wheel fully aft during the landing rollout.

____ **Q 202**

True or False? A pilot is maintaining altitude at the absolute ceiling of his airplane. Therefore, the airplane is on the verge of stalling.

____ **Q 203**

The dizzying effects of an inverted spin make it difficult to quickly determine which rudder to apply when attempting recovery. Therefore, the pilot should apply rudder
    A. opposite to the turn needle.
    B. opposite to the turn coordinator.
    C. opposite to the slip/skid ball.
    D. toward the slip/skid ball.
    E. None of the above.

A pilot needs to make a takeoff in the shortest distance possible. He should load the airplane so that it has

  A. an aft center of gravity.
  B. a mid center of gravity.
  C. a forward center of gravity.
  D. It does not matter.

Normal approach speed is determined by multiplying the ________________ stall speed in the landing configuration by ______ and then converting the result to ________________ airspeed.

  A. indicated, 1.2, calibrated
  B. calibrated, 1.2, indicated
  C. indicated, 1.3, calibrated
  D. calibrated, 1.3, indicated

A pilot knows the maximum cruise speed and climb performance of his airplane when flown at its maximum allowable gross weight. He is curious, however, to know what the performance of his aircraft would be at twice this weight. How can he determine this during flight without overloading the airplane?

True or False? Pilots are aware of adverse yaw effect created by ailerons. Many airplanes also exhibit a characteristic known as adverse roll.

True or False? An inverted aircraft in cruising flight receives the greatest stress from an updraft (as compared to an equally strong downdraft).

What is wrong with this statement: A pilot is cruising at 17,500 feet, hears *The High and the Mighty* on his low/medium-frequency receiver, removes his oxygen mask for a moment, sips water from his thermos bottle, and then whistles along with the music.

During flight, the pilot of a Cessna 172 is shocked to discover that the up-elevator cable of the control system has snapped. How can he make a safe landing using the control wheel?

A pilot stalls his conventional lightplane, kicks and holds full rudder, pauses for a few seconds, adds full power, and then applies and holds full opposite aileron. The resultant maneuver is called a ______________.

Why does an airplane with a constant-speed propeller have a shorter takeoff roll than an identical airplane with a fixed-pitch propeller (everything else being equal)?

During a hard landing, the wings of an airplane tend to flex downward. At such a time, a G-meter in the airplane indicates

    A. more than +1 G.
    B. between 0 and +1 G.
    C. between 0 and -1 G.
    D. more than -1 G.

The most direct and adverse effect of humidity on aircraft performance is

    A. reduced lift.
    B. reduced propeller thrust.
    C. reduced horsepower.
    D. increased drag.

How is it possible for the pilot of an airplane to totally eliminate induced drag and wingtip vortices while in flight?

Which of the following does not belong?
   A. decreased cruise speed.
   B. decreased stall speed.
   C. easier to increase G loading.
   D. reduced longitudinal stability.

A pilot is making a crosswind landing in a tricycle-gear airplane. If the airplane is allowed to touch down on just the main gear while in a wings-level crab, it would initially tend to
   A. turn into the crosswind.
   B. turn downwind.
   C. travel in the direction in which it is headed.
   D. straighten out and go in the direction in which it was traveling over the ground immediately before touchdown.

A pilot is gliding at the optimum indicated glide speed and then enters a turn. To maintain a maximum-range glide during the turn, glide speed should
   A. be increased.
   B. be decreased.
   C. remain the same.
   D. Cannot be determined.

Explain the circumstances under which it is possible for a sailplane pilot to maintain an approximately constant altitude for extended periods of time without the presence of vertical air currents.

True or False? A pilot practicing steep turns in smooth air encounters a short, turbulent bump at the completion of 360 degrees of turn. This means that he has maintained altitude during the turn.

True or False? A pilot is executing an instrument approach to an airport obscured by dense fog. The error resulting from using an incorrect altimeter setting that is too low is more dangerous than using an incorrect altimeter setting that is too high.

A pilot is flying directly over an airport that has an elevation of 5,000 feet at an indicated altitude of 8,000 feet MSL while using an updated altimeter setting of 30.42 inches Hg. The air temperature from the ground up to the pilot's altitude is only 1-degree Celsius warmer than standard. Without using a computer, arrange the following alphabetically listed altitudes in order of magnitude (lowest altitude first and highest altitude last):

   A. absolute altitude
   B. density altitude
   C. indicated altitude
   D. pressure altitude
   E. true altitude

True or False? A passenger is seated behind the center of gravity of an airplane and is holding a slip-skid instrument in his hand, so that he can observe it while facing forward. The pilot depresses the right rudder pedal. At such a time, the ball in the pilot's instrument moves left, but the passenger's ball moves temporarily toward the right side of the airplane.

True or False? Pilots operating conventional piston engines with constant-speed propellers can save fuel and reduce engine wear by reducing rpm and increasing manifold pressure to maintain the same horsepower as is obtained with increased rpm and reduced manifold pressure.

A pilot is leveling off after climbing to cruise altitude. As airspeed increases, he notices that nose-up trim is required to maintain altitude. Assuming that the trim tab is operating normally, what is the most likely cause of the problem?

How is it possible to turn a Cessna 172 while using a 90-degree bank angle without overstressing the airplane?

The static ports of an airplane become clogged with ice while at cruise altitude. During descent in continuing icing conditions, indicated airspeed is _____________ than it normally would be, and indicated altitude is _____________ than it normally would be.

   A. higher, higher
   B. lower, lower
   C. higher, lower
   D. lower, higher

If indicated rate of climb or descent is multiplied by 10, vertical speed is expressed (approximately) in

   A. kilometers per hour.
   B. knots.
   C. miles per hour.
   D. None of the above.

A non-instrument-rated pilot needs to descend through a layer of cloud without the assistance of any gyroscopic instruments and while referring only to a conventional compass to keep his wings level. The best magnetic heading to use for this purpose is

    A. north.
    B. east.
    C. south.
    D. west.

True or False? If the nosewheel tire of a typical single-engine airplane is flat and the nosewheel strut is bottomed (it contains no air or hydraulic fluid), a rotating, two-bladed propeller will scrape the ground.

When the surface wind blows across a runway at a _______-degree angle, the crosswind component during takeoff or landing is exactly half of the reported wind speed.

How is it possible to make a 360-degree turn such that the magnetic compass and gyroscopic heading indicator agree throughout the turn (no compass-turning error)?

Structural icing on the leading edges of horizontal stabilizers typically causes an airplane to

    A. pitch down.
    B. pitch up.
    C. roll left.
    D. pitch up and roll left.

A pilot arrives over his destination only to find the single runway closed because of a disabled aircraft. To loiter above the airport for as long as possible before having to divert to an alternate airport, the pilot should maintain altitude using

A. V_X.
B. V_Y.
C. the speed for best glide.
D. None of the above.

True or False? Structural icing typically forms on the tail before forming on the wings.

True or False? Two identical Cessna 172s — one heavy and one light — encounter engine failure at the same altitude and are subsequently flown at the best glide speed as published in the pilot's operating handbook. The heavier airplane will glide farthest.

A pilot in cruise flight with a fuel-injected engine notices a simultaneous onset of engine roughness and excessive fuel pressure. What should be done to resolve the problem?

True or False? While performing a pre-takeoff magneto check using a conventional key-type, rotary ignition switch, a pilot observes noticeable engine roughness with the switch in the "R" position. The problem involves the left magneto.

True or False? Shortly after takeoff in a Cessna 337 Skymaster, the pilot experiences failure of the rear engine. This results in noticeably poorer climb performance than if the front engine had failed.

How is it possible in many types of airplanes to experience reversed effect of a primary flight control during upright, 1-G flight?

Which of the following does not belong?
- A. aileron
- B. barrel
- C. slow
- D. snap

True or False? Pressing the balls of both feet firmly and equally against the bottoms of both rudder pedals reduces aircraft yaw in turbulence.

Why does it become progressively easier to overstress an airplane as the center of gravity moves aft?

True or False? During a climb in icing conditions, the pitot tube becomes obstructed with ice. As the climb progresses, indicated airspeed becomes increasingly less than would normally be expected.

True or False? The typical general aviation altimeter cannot be set using an altimeter settling in excess of 31.00 inches of Mercury.

What is the most common reason for gear-up landings?

_______________________________________ **Q 247**

A Raytheon B36-TC Bonanza and a Piper Malibu Mirage have wing areas of 187.8 and 183.8 square feet, respectively, and maximum-allowable gross weights of 3,850 and 4,300 pounds, respectively. Everything else being equal, which airplane experiences the greatest G-load in response to a given vertical gust and why?

_______________________________________ **Q 248**

Two pilots flying in formation with inaccurate tachometers want to fly at exactly the same rpm. How can they do this?

_______________________________________ **Q 249**

A Cessna 172 has only one static port, which is on the left side of the fuselage. If a forward slip is performed at a given indicated airspeed, which is safer, a left or right slip?

_______________________________________ **Q 250**

True or False? A pilot flying a typical single with a fixed-pitch propeller enters a dive and accelerates to V_{NE} with the throttle fully closed. He should anticipate that engine rpm will exceed the redline.

_______________________________________ **Q 251**

Name two ways in which an airplane might be descending without its pilot being aware of it.

_______________________________________ **Q 252**

A pilot performs a wings-level, power-off stall in a typical single with the gear and flaps retracted. He notes that the indicated stall speed is below normal. One reason for this might be that

    A. the center of gravity is unusually far forward.
    B. the center of gravity is unusually far aft.
    C. one of the two static ports is clogged.
    D. aircraft gross weight is excessive.

True or False? A pilot is descending through 1,000 feet AGL while executing an instrument approach in dense cloud during the migration season. He should be concerned about the possibility of a bird strike.

True or False? A pilot is making a visual, straight-in approach to a runway with a pronounced upslope. He will tend to overshoot the normal touchdown zone.

True or False? A tetrahedron is a reliable source of wind direction.

When scud running, why is it that a pilot often is deceived into believing that the weather ahead of the airplane usually is better than the weather behind?

A pilot is changing heading from north to northwest (in the shortest direction) while performing an inverted turn. To initiate recovery from this turn he should simultaneously move the

    A. right aileron up and apply right rudder.
    B. left aileron up and apply right rudder.
    C. right aileron up and apply left rudder.
    D. left aileron up and apply left rudder.

Three-time U.S. National Aerobatic Champion Patty Wagstaff. (Q257)

Questions

$$\overline{\hspace{7cm}}\ \mathbf{Q\,258}$$

True or False? Using observed indicated airspeed to obtain calibrated airspeed and then correcting the result for temperature and pressure altitude results in true airspeed.

$$\overline{\hspace{7cm}}\ \mathbf{Q\,259}$$

At sea level, V_Y of a typical, normally aspirated (nonturbocharged) single is greater than V_X and decreases with altitude. Similarly, V_X increases with altitude. V_Y and V_X become the same at

______________.

$$\overline{\hspace{7cm}}\ \mathbf{Q\,260}$$

True or False? In a given airplane, maneuvering speed (V_A) decreases as gross weight increases.

$$\overline{\hspace{7cm}}\ \mathbf{Q\,261}$$

A Lockheed SR-71 Blackbird is in a 30-degree banked turn at 85,000 feet and a true airspeed of 1,800 knots. The circumference of the resulting circle is

    A. 215 NM.
    B. 515 NM.
    C. 815 NM.
    D. 1,115 NM.

$$\overline{\hspace{7cm}}\ \mathbf{Q\,262}$$

True or False? A tricycle-gear airplane with a 50-knot stall speed is tied down loosely while pointed into a 60-knot wind. The wings cannot develop sufficient lift to cause the airplane to rise off the ground.

$$\overline{\hspace{7cm}}\ \mathbf{Q\,263}$$

True or False? A typical single with retractable landing gear is descending on a typical glide path (or VASI) at its best glide speed in the landing configuration. The engine fails when 2 miles from the airport. By maintaining the best glide speed and immediately retracting the gear and flaps, the pilot most likely can glide to a safe landing on the runway.

A pilot is about to fly a typical, piston-powered, lightplane that was manufactured in 1975. He should know that the airspeed indicator most likely reads in _______ and the color markings represent _____________ airspeed.

 A. mph, calibrated
 B. mph, indicated
 C. knots, calibrated
 D. knots, indicated

A pilot is in a full-power, vertical dive toward a target directly beneath the aircraft. What must be the pitch attitude of the aircraft?

True or False? Baggage-compartment doors on most general aviation airplanes should be latched and locked by key during flight.

A pilot notices during a VFR climb that the airspeed indicator, VSI, and altimeter suddenly and simultaneously become erratic. And then, just as suddenly, they revert to seemingly normal operation. What is the most likely problem?

True or False? Rain falling on a windshield creates an optical illusion that leads a pilot to believe that he is higher during a landing approach than he really is.

Why does a typical general aviation airplane pitch down during a stall entry?

A pilot is performing an on-pylon (not an around-pylon) turn and notices that the pylon is moving behind the wingtip. He should

   A. climb and increase airspeed.
   B. climb and decrease airspeed.
   C. descend and increase airspeed.
   D. descend and decrease airspeed.

True or False? A pilot about to land on an unusually wide runway at night is most likely to perceive that he is closer to the runway than he really is and flare prematurely (while too high).

True or False? A lightplane pilot on final approach is number two to land following a Boeing 747. If he could see the wingtip vortices, he would notice that the vortex generated by the right wingtip of the 747 is rotating in a clockwise direction.

A pilot wants to fly as perfectly a rectangular pattern as possible while in the left traffic pattern to Runway 36 at a time when the wind is strong and from the northeast. This requires that the radius of all corners be the same with respect to the ground. Assuming a constant airspeed throughout the pattern, the most steeply banked turn will be required when turning from

   A. base to final.
   B. crosswind to downwind.
   C. downwind to base.
   D. upwind to crosswind.

"Lights! Camera! Action!" is an expression frequently heard on the set of a Hollywood movie studio. How does it apply to pilots?

A high-altitude airplane is climbing at an indicated airspeed of 120 knots, a true airspeed of 180 knots, and a vertical speed of 600 fpm. The resultant climb gradient is

A. 200 feet per NM.
B. 250 feet per NM.
C. 300 feet per NM.
D. Cannot be answered with the information provided.

True or False? During an approach to land an amphibian on water with the landing gear extended, the green landing-gear light(s) and the red gear-warning light will be illuminated.

A pilot performs a 90-degree climbing left turn during which airspeed steadily decreases to near stall, and bank angle increases steadily. The aircraft is then made to transition into a 90-degree descending left turn during which airspeed steadily increases and bank angle steadily decreases. This maneuver is called a

__________________.

Generally speaking, which types of piston engines should be leaned for takeoff from airports at high density altitudes and which should not?

Why is the application of carburetor heat least effective in preventing carburetor ice during a gliding approach in an airplane, a time when it is needed the most?

A pilot is on a long straight-in approach to a runway at night. During his descent, the tower controller very gradually turns down the intensity of the runway lights. The most likely result is that the pilot would unwittingly

    A. descend below the normal descent profile.
    B. rise above the normal descent profile.
    C. The descent profile would not be affected.
    D. The result cannot be determined.

True or False? Some bush pilots use a procedure of landing their wheel-equipped taildraggers on the surface of a deep lake (without floats or skis). After landing on the main gear, they decelerate to below stall speed, allow the tailwheel to touch down while still on the surface of the lake, and then roll onto a landing area along the shore that otherwise would be too short for a conventional landing.

When encountering light turbulence, indicated airspeed varies between 5 and 15 knots. What airspeed variation would a pilot expect to see during an encounter with extreme turbulence?

A pilot is taxiing on a steady heading of 315 degrees under the influence of a brisk wind. He is holding the left aileron down and the control wheel fully aft. Both of these control inputs are opposite to what is correct because the wind direction is

    A. northerly.
    B. easterly.
    C. southerly
    D. westerly.

An aircraft generates the strongest, most-threatening wake vortices when it is

A. fast, heavy, and clean (flaps retracted).
B. fast, heavy, and dirty (flaps extended).
C. slow, heavy, and clean (flaps retracted).
D. slow, heavy, and dirty (flaps extended).

True or False? The pilot of a Cessna 172 is executing an ILS approach at the minimum-approved distance behind a Boeing 747-400. Both aircraft are stabilized on the glide slope. The Cessna pilot can expect to encounter wake turbulence.

What is the difference between an outside loop and an inverted outside loop?

Which of the following does not belong?

A. chandelle
B. Immelmann turn
C. lazy eight
D. wingover

Which of the following does not belong?

A. accelerated
B. departure
C. hammerhead
D. whip

When flying new or modified airplanes, test pilots occasionally enter spins (intentionally or otherwise) from which recovery using the flight controls may not be possible. How do they recover from such spins?

Q 290

A pilot enters a loop. After completing slightly more than one-fourth of the maneuver, the airplane is in a near-vertical, slightly inverted climb. At this point, the pilot rapidly applies full forward pressure to the control stick while simultaneously applying maximum and opposite aileron and rudder. The resultant aerobatic maneuver is called a ______________________.

Q 291

A cliché states that a good landing is one from which a pilot can walk away. What is the definition of a great landing?

Q 292

Following an engine failure at night, the pilot of a single-engine airplane turns on the landing light as he glides toward the ground but doesn't like what he sees. What should he do?

Q 293

The actual magnetic heading of an aircraft in a stabilized inverted turn is increasing. (When viewed from above, the aircraft is turning in a clockwise direction.) During this turn, the conventional (wet) compass

    A. indicates a constant heading.
    B. indicates a steady decrease in magnetic heading.
    C. indicates a steady increase in magnetic heading.
    D. is unpredictable.

Q 294

A pilot opens the cockpit door, lowers the flaps, turns on the carburetor heat, retards the throttle fully, and turns off a magneto. Why is he doing these things?

Which of the following does not belong?
  A. inability to arrest descent rate
  B. lack of pitch authority
  C. lack of roll control
  D. lack of yaw control

True or False? During cruise flight, a pilot experiences total power failure in his single-engine airplane. To maximize glide range, he should pull up the nose and convert excess airspeed into altitude (until reaching the best glide speed).

What is the V-speed that represents the minimum indicated airspeed at which an airplane can be made to lift from a runway?

While decelerating on the runway after landing, the reverse thrust used to assist in slowing jet aircraft typically is reduced or cancelled at an indicated airspeed of approximately 60-80 knots. Why is reverse thrust not normally used below this speed?

A pilot initiates a normal glide at 10,000 feet MSL and maintains the published (best) glide speed throughout his descent. As altitude decreases, the rate of descent will ___________ and the glide angle (or ratio) will ____________.
  A. increase, increase
  B. increase, decrease
  C. decrease, increase
  D. decrease, decrease
  E. None of the above.

_______________________________________ **Q 300**

True or False? Landing on a runway coated with wet ice requires
more distance than landing on a runway coated with ice that is dry.

_______________________________________ **Q 301**

Many unimproved airports have "runway halfway" signs that
indicate to pilots the midpoint of a runway. According to the
Aeronautical Information Manual, what rule of thumb can a
departing pilot use to enhance safety at such an airport?

_______________________________________ **Q 302**

True or False? Spin recovery in a typical single-engine airplane
includes reducing the angle of attack, applying antispin rudder,
neutralizing the rudder, and recovering from the dive (in that
sequence).

_______________________________________ **Q 303**

A pilot flying over California is maintaining a magnetic heading
of 180 degrees. If he were to begin rolling into a left turn, his
magnetic compass would initially indicate a turn toward the
________. If he were to begin rolling into a left turn while on a
magnetic heading of 360 degrees, his compass would initially
indicate a turn toward the ________.

   A. east, east
   B. east, west
   C. west, east
   D. west, west

_______________________________________ **Q 304**

What is the difference between an inverted normal loop and an
inverted outside loop?

_______________________________________ **Q 305**

What is the easiest way to fly an airplane at a constant angle
of attack?

Because of a mechanical problem, a pilot flying a retractable-gear airplane has to make a gear-up landing. In most cases, his best choice of landing sites would be a

A. dirt runway.
B. foamed, hard-surface runway.
C. grass runway.
D. hard-surface runway that has not been foamed.

Why do pilots of airplanes with radial engines turn the propellers by hand before the first flight of the day?

True or False? A pilot is about to land an airplane with the right landing gear and nosewheel extended but with the left gear stuck in its well. The ailerons are equipped with conventional trim tabs that are controllable from the cockpit. During landing and rollout, the right aileron tab should be deflected fully down.

Provide two reasons why it usually is more advantageous during VFR flight to lower the nose and increase airspeed when flying through a downdraft than it is to add power and attempt to maintain altitude.

True. Furthermore, if the pilot allows the stall to deepen, the horizontal tail will stall, which causes the nose to drop.

_______________________________________ A164

The airplane climbing within the low has the best climb performance. This is because air within a high subsides and air within a low rises. An airplane obviously climbs better in rising air than it does in subsiding (or sinking) air.

_______________________________________ A165

D. The northerly wind causes groundspeed to be greatest on the west side of the circle. Consequently, the least amount of time is available to turn a given amount of arc on that side of the circle. This necessitates a steeper bank angle and greater turn rate than elsewhere on the circle.

_______________________________________ A166

B. An airplane is an oversized weather vane and tends to turn into the wind. In an updraft, the relative wind comes from below. Upon entering the updraft, therefore, the airplane tends to turn into the vertically rising air, which means that it will tend to pitch nose down.

_______________________________________ A167

A. After the glass is broken, air flows backwards through the vertical-speed indicator. The amount of indicated vertical speed is correct, but its direction is reversed. This is why a climb is indicated as a descent, and vice versa.

$$A168$$

C. This is difficult to visualize but is nevertheless true. It helps to consider that an airplane in a 90-degree banked turn at constant altitude is only pitching. An airplane in a normal turn at a constant altitude is yawing and pitching. Beginning a climb or descent while turning induces the rolling motion.

$$A169$$

D. Steepening bank angle increases turn rate, but increasing airspeed decreases turn rate. Without knowing how much the bank angle and airspeed are increased, it is impossible to know whether turn rate increases or decreases.

$$A170$$

B. Strange but true. To maintain the same glide ratio, glide speed must be reduced as aircraft gross weight decreases. If glide speed is not reduced, glide performance and glide ratio decrease as the aircraft gets lighter. Conversely, an airplane that gains weight must be flown at an increased airspeed to retain the same glide performance.

$$A171$$

A. There normally is no change in static-system pressure during takeoff because the aircraft does not change altitude. This is why all pitot-static instruments would operate normally during the takeoff roll. During climb, however, indicated airspeed would become less than normal. This error would increase with altitude.

$$A172$$

B. Temperature changes do not affect an altimeter on the ground (or very close to it). Altimeter errors due to nonstandard temperature variations increase with altitude above the elevation of the nearest ground station from which the altimeter setting is available.

A. The climb rate of a lightplane with a normally aspirated (non-turbocharged) engine decreases in linear fashion. In this case, the climb rate decreases 100 fpm during each thousand feet of climb. Since the service ceiling of a single-engine airplane is the highest altitude at which it can climb 100 fpm, the service ceiling of this airplane must be 14,000 feet. Its absolute ceiling is 15,000 feet.

B. The region of reversed command is the formal name given to that mode of flight where the airplane is flying behind the power curve. Additional power is required to maintain altitude as airspeed decreases.

D. V_X increases and V_Y decreases such that they eventually become the same at the airplane's absolute ceiling.

B. During a right skid, some ram air pressure enters the left static source. This normally is countered by reduced pressure sensed by the right static source on the "downwind" side of the airplane. But when the right source is blocked, ram air entering the left source "pressurizes" the static system somewhat, which causes the instruments to sense and "believe" that atmospheric pressure has increased.

A. and D. The spiraling slipstream pushes against the left side of the vertical fin, which causes a left yaw. Torque causes the aircraft to tend to roll left, which increases the force on the left tire. This increases the rolling friction created by that tire and, therefore, a tendency to yaw left.

$$A 178$$

True. Just as stall speed increases with an increase in G loading, it decreases with a decrease in G loading. At zero Gs, an airplane has no stalling speed and cannot be made to stall.

$$A 179$$

A. Shifting an airplane's center of gravity aft causes cruise speed to increase and stall speed to decrease, and vice versa.

$$A 180$$

C. Pushing the stick forward raises the nose with respect to the horizon and results in a climb. The stick is moved in the direction of the desired turn, which — in this case — is left.

$$A 181$$

C. The erecting mechanism of the gyro very slowly reacts to the apparent force of gravity (which — in a turn — is perpendicular to the lateral axis of the airplane) and gradually reduces the indicated bank angle. This error is eliminated shortly after the pilot rolls out of the turn.

$$A 182$$

B. Although the indicated liftoff airspeed at the higher altitude is the same as at sea level, this translates into a higher ground-speed, which increases the required takeoff distance.

$$A 183$$

D. A knife-edge turn is effectively a loop performed on a horizontal plane. Because the airplane is only pitching and not yawing, the turn indicator does not sense yaw and, therefore, shows a zero turn rate.

_______________________________________ A184

A. A flame is buoyant and rises opposite to the force of gravity. During acceleration, the apparent force of gravity is from below and behind (which is why a pilot is pressed into both the bottom and the back of his seat). The flame, therefore, both rises and leans forward. You can observe this phenomenon in an automobile, but please do not try it in an airplane.

_______________________________________ A185

B. At the top of a positive-G loop, the force of gravity is perceived to come from the opposite direction, which is why the pilot stays firmly in his seat while inverted. Because the flame burns opposite to the perceived force of gravity (see the answer to the previous question), the flame burns (points) earthward, or toward the ceiling of the airplane.

_______________________________________ A186

C. Reducing airspeed reduces sink rate, which causes the airplane to remain in the air longer. This increases the time during which the tailwind can assist the pilot in extending his glide range. The amount of airspeed reduction depends on wind speed.

_______________________________________ A187

A. A spinner smoothly diverts the oncoming airflow into the cowling inlets, which optimizes engine cooling.

_______________________________________ A188

A. The application of carburetor heat allows induction air to bypass the air filter. This allows unfiltered air containing dirt and other contaminants from the ground to enter the engine and possibly cause damage.

A189

Each pilot watches the shadows of their wingtips. If the shadows do not touch, then neither can the wings. Using shadows also can be a good way to observe rudder movement during the preflight runup when the airplane does not have a rear window.

A190

False. The angle of attack is not sufficiently large to cause the wings to stall. Although the aircraft does not have sufficient speed to fly, its wings are nevertheless developing some lift. An attempt to lift off at 40 knots, however, would cause a stall because of the enlarged angle of attack.

A191

True. A windmilling propeller causes the engine to turn and ambient atmospheric pressure is allowed to fill the cylinders as long as the throttle is open. Closing the throttle would reduce manifold pressure.

A192

75 degrees

A193

An increase in rpm indicates that the mixture is too lean. Applying carburetor heat enriches the mixture and, therefore, would cause power to increase.

A194

B. During a normal glide, the glide ratio (ratio of forward speed to vertical speed) remains constant. As the aircraft descends at a constant indicated airspeed, its true airspeed decreases, which means that the vertical speed must decrease in the same proportion.

____ **A**195

False. Airspeed has nothing to do with load factor in a stabilized turn. Whether descending, climbing, or in level flight, load factor for a given bank angle is the same.

____ **A**196

C. 60 degrees

____ **A**197

A snap roll, which the British refer to as a flick roll.

____ **A**198

C. When there is a headwind on the base leg, there also is a crosswind on both the downwind and final-approach legs. This necessitates crabbing, which reduces groundspeed and glide performance.

____ **A**199

Focus the eyes on one object on the horizon for as long as possible, then shift to another, and so forth.

____ **A**200

True. The aircraft first encounters the outflow portion of the microburst, which results in an increasing-headwind type of wind shear. Powerful downdrafts and an increasing-tailwind type of shear soon follow.

____ **A**201

True. Nose-down trim (moving the tab up) increases elevator effectiveness but requires more effort to hold the control wheel aft.

$$A202$$

False. The only airspeed that allows a pilot to maintain this altitude is either V_X or V_Y (which are the same at the absolute ceiling). Any other airspeed would result in descent.

$$A203$$

A. The turn coordinator and the slip-skid ball do not provide reliable recovery guidance during an inverted spin.

$$A204$$

A. An aft center of gravity reduces stall speed, which reduces liftoff speed, which reduces takeoff distance.

$$A205$$

D. Multiply calibrated stall speed by 1.3 and then convert the result to obtain the indicated approach speed.

$$A206$$

The pilot simply notes the performance of his aircraft while stabilized in a 60-degree banked turn. At such a time, the aircraft experiences a 2-G load factor and effectively weighs twice as much as when the wings are level in stabilized flight.

$$A207$$

True. When the rudder is deflected, a horizontal force is created by the vertical stabilizer at a point that usually is above the longitudinal axis of the airplane. This force initially tends to roll the aircraft in the opposite direction.

$$A208$$

True. It does not matter whether the aircraft is inverted or not.

The reduced air density at 12,000 feet makes whistling difficult. At higher altitudes, it becomes impossible. Try it.

He applies substantial nose-up trim and applies forward pressure as necessary on the control wheel to maintain the desired attitude. He flares by gradually releasing most or all of this forward pressure.

A flat spin from which recovery might not be possible.

An engine with a constant-speed propeller develops 100-percent power from a standing start; an engine with a fixed-pitch propeller does not.

Fixed-pitch propeller on a Luscombe 8E.   (A212)

-- **A213**

A. Similarly, the pilot is forced harder into his seat, which indicates more than +1 G.

-- **A214**

C. Water vapor displaces some of the dry air that would otherwise be available for combustion, enriches the mixture, and retards flame propagation within the cylinders.

-- **A215**

By pitching down and establishing a zero-G condition, the wings cannot develop lift, induced drag, or wingtip vortices.

-- **A216**

A. The other choices are the consequences of an aft center of gravity.

-- **A217**

D. A tailwheel-equipped airplane would not be quite so forgiving.

-- **A218**

A. Turning flight results in an increased load factor, which increases the effective weight of the airplane.

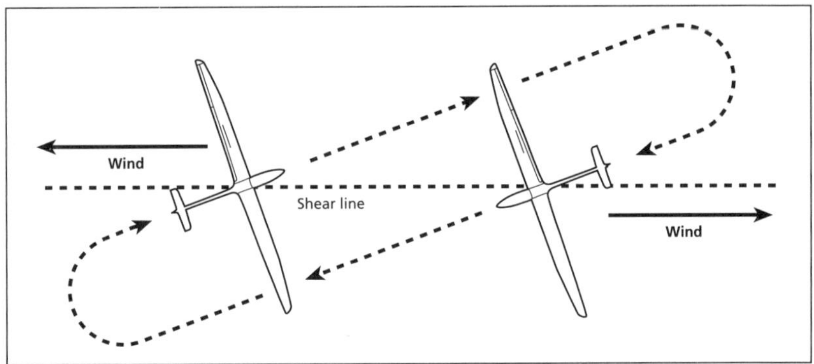

Use in conjunction with Answer 219.

$$A_219$$

The glider crosses a wind-shear line (*see* figure above), resulting in an airspeed gain, which is converted to altitude. The pilot turns right to re-cross the shear line, resulting in another airspeed gain, and so forth. The albatross uses similar techniques of dynamic soaring to fly long distances over water.

$$A_220$$

False. Wake turbulence descends behind an airplane. The bump felt by the pilot was the result of flying through his own wake at some point below the altitude at which the wake was generated.

$$A_221$$

False. When using an altimeter setting that is too low, the altimeter indicates a lower-than-actual altitude. In other words, the aircraft is above the indicated altitude.

$$A_222$$

A. Absolute altitude = 3,000 feet (above the ground).
D. Pressure altitude = 7,500 feet (8,000 feet less 500 feet).
B. Density altitude = slightly more than 7,500 feet.
C. Indicated altitude = 8,000 feet (given).
E. True altitude = slightly more than 8,000 feet.

True. The nose moves to the right at such a time, which causes a normal slip-skid ball to move left. The tail, however, moves left, which causes a slip-skid ball behind the CG to move right. A slip-skid ball situated at the pivot point of the aircraft would not move at all during a wings-level skid. This explains why the most comfortable ride in an airplane is in a seat near the aircraft's center of gravity.

True. Most modern piston engines have an approved power range for which this practice is approved. Often, however, the approved power settings are not published in the pilot's operating handbook for the airplane and must be obtained from the operator's handbook published by the engine manufacturer.

The airplane is longitudinally unstable, which most likely is the result of an excessively aft center of gravity.

Do not attempt to maintain altitude. Note, however, that banking in excess of 60 degrees would result in an unauthorized aerobatic maneuver.

A. Indicated airspeed is excessive because ram air pressure increases within the instrument during descent, while static air pressure does not. The altimeter continues to indicate cruise altitude, because the instrument does not sense a change in static pressure.

_______________________________ A 228

B. A VSI indicates vertical speed in thousands of feet per minute. For example, 6,000 fpm is shown as "6." 6,000 fpm = approximately 1 NM per minute, which equals 60 knots. Multiplying "6" by 10 also results in 60.

_______________________________ A 229

C. On easterly and westerly headings, the compass errs because of pitch and airspeed changes. On northerly and southerly headings, it errs because of turning errors. When heading south, however, the compass moves in the proper direction as soon as the airplane begins to bank, which makes it the most suitable heading for maintaining a wings-level attitude.

_______________________________ A 230

False. The propellers of normal-, aerobatic-, and utility-category airplanes will clear the ground. Also, when all tires are inflated and the struts properly extended, the propeller will clear the ground by a minimum of 7 inches (according to 14 CFR Part 23 certification requirements).

_______________________________ A 231

30 degrees. In trigonometry, the sine of 30 degrees is 0.5.

_______________________________ A 232

Execute a wings-level skidding turn at a constant airspeed. Compass turning error is the result of banking. This can be demonstrated during a turn made while taxiing.

_______________________________ A 233

A. The normal function of a conventional horizontal stabilizer is to produce a downward force that prevents the nose from dropping. Ice interferes with this function.

_______________________________ **A234**

D. The pilot should use the speed for maximum endurance,
which is normally not provided in pilot operating handbooks.
It is the speed that results from using the minimum power
required to maintain a given altitude and can be determined
by trial and error.

_______________________________ **A235**

True. Structural icing typically forms first on objects with a
small radius (such as an outside air-temperature probe). The
leading edges of tail surfaces usually have smaller radii than
the leading edges of wings.

_______________________________ **A236**

True. The best glide speed applies when the aircraft is loaded to
its maximum-allowable gross weight. A lightly-loaded airplane
must be flown at slightly less airspeed to achieve the same glide
performance.

_______________________________ **A237**

In most cases, the fuel-air mixture should be leaned until the
roughness disappears.

_______________________________ **A238**

False. With the key in the "R" position, the engine is operating
on the right magneto and is the source of the problem.

_______________________________ **A239**

True. The rear engine of a Skymaster provides more thrust than the
front engine. This presumably is because the fuselage does not interfere with the air being thrust rearward by the pusher (aft) propeller.

A240

Aileron reversal can occur when attempting to roll while on the verge of a stall. An aileron deflected downward (to raise a wing) increases the angle of attack of that wing, induces a stall, and lowers the wing.

A241

D. The snap roll is essentially a spin performed about a horizontal axis and is the only one of the four rolls performed while the airplane is stalled.

A242

True. During a right yaw, for example, the tail moves left and the rudder would otherwise move right, because the relative wind (from the left) would push it that way. This exacerbates yaw. Holding the rudder in place improves directional stability.

A243

The pulling force on the control wheel required to create a given G-load becomes less as the center of gravity moves aft. At an extremely aft center of gravity, excess Gs can be "pulled" using only the fingertips.

A244

False. The airspeed indicator will behave like an altimeter. As altitude is gained, indicated airspeed becomes increasingly greater than would normally be expected.

A245

True. Also, such altimeters typically cannot accept a setting below 28.10 inches of Mercury. Such limitations can present operational difficulties during periods of extremely high or low atmospheric pressure.

_______________________________________ **A246**

Water landings in amphibians. (Sorry about that.)

_______________________________________ **A247**

The B36TC and the PA-46 have wing loadings (gross weight divided by wing area) of 20.5 and 23.4 pounds per square foot, respectively. The B36TC has the lighter wing loading, which results in the most acceleration (G-load).

_______________________________________ **A248**

The wingman moves behind the leader and views the leader's propeller disc through his own. Because of stroboscopic effect, the wingman will see a line (assuming that the propellers of each aircraft have the same number of blades). He adjusts his throttle until the line remains stationary.

_______________________________________ **A249**

During a left slip, ram air from the left enters the static port and pressurizes the static line to the airspeed indicator. This reduces IAS and is a safe-side error. During a right slip, IAS is greater than what it should be and is not as safe.

_______________________________________ **A250**

True. Aircraft certification regulations allow up to 10-percent overspeed under these conditions. This should be avoided, however, as propeller and engine damage could eventually occur.

_______________________________________ **A251**

When flying into colder air and lowering atmospheric pressure. ("Cold and low, look out below.")

_______________________________________ **A252**

B. Stall speed reduces as the center of gravity moves aft and increases as the center of gravity moves forward.

False. According to the Audubon Society, birds are VFR creatures and do not fly in clouds. Like some pilots, however, small birds can get "caught" in IFR conditions and fly into tall buildings.

False. He will tend to undershoot. A pronounced downslope most often leads to an overshoot.

False. Tetrahedrons often are locked into position to designate landing direction, not wind direction.

When looking aft, ground objects disappear in the veil of reduced visibility, creating the illusion of worsening visibility. When looking ahead, conditions seem to be improving because forward motion causes progressively more terrain to come into view.

C. The airplane must roll clockwise about its longitudinal axis (as viewed from behind). Aileron input is the same, therefore, as when recovering from an upright left turn. The rudder, however, must be moved in the opposite direction (to the left) to compensate for adverse yaw effect.

False. This fails to consider instrument error, which varies between individual airplanes of the same make and model.

The airplane's absolute ceiling.

$$A260$$

False. V_A decreases as gross weight decreases.

$$A261$$

B. The turn radius is 82 NM. The Blackbird holds the current world speed record for jet airplanes of 1,907 knots, but many believe that it can fly faster.

$$A262$$

True. The angle of attack of the wings is too small (and possibly negative) to generate sufficient lift.

$$A263$$

False. To glide along at 3-degree glide slope or VASI requires an airplane to have a 20:1 glide ratio; the typical single can glide only half that well.

$$A264$$

A. This generally applies to aircraft manufactured prior to or during 1975. Answer D is generally correct for aircraft manufactured after 1975.

$$A265$$

More than 90 degrees nose-down (slightly inverted) so that the horizontal component of thrust exactly equals and counteracts the net horizontal component of lift. If a 90-degree, nose-down attitude is maintained, the horizontal component of wing lift would cause the aircraft to drift "ahead" of the target.

$$A266$$

False. Locking a properly latched baggage-compartment door usually does not close the door any more securely, but it can hamper escape or ground-rescue operations should this become necessary.

$$\underline{\qquad\qquad\qquad\qquad\qquad\qquad\qquad\qquad} A267$$

Water in the static line. The erratic indications are caused by air bubbling through the trapped water.

$$\underline{\qquad\qquad\qquad\qquad\qquad\qquad\qquad\qquad} A268$$

False. Moisture on a windscreen produces unpredictable illusory effects because of the irregular refraction of light caused by raindrops. Depending on moisture pattern and windscreen shape, a pilot might perceive being higher or lower than he really is.

$$\underline{\qquad\qquad\qquad\qquad\qquad\qquad\qquad\qquad} A269$$

A downward aerodynamic force on the tail normally keeps the tail down and the nose up. During a stall, air no longer flows smoothly from the inboard sections of the wings to the tail. This causes a conventional tail to lose effectiveness, which allows the nose to pitch down.

$$\underline{\qquad\qquad\qquad\qquad\qquad\qquad\qquad\qquad} A270$$

B. When the pylon slips behind the wingtip, the aircraft is below its pivotal altitude for the airspeed being used. The pilot should gain altitude, reduce airspeed, or both.

$$\underline{\qquad\qquad\qquad\qquad\qquad\qquad\qquad\qquad} A271$$

True. Conversely, a pilot about to land on an unusually narrow runway can perceive that he is higher than he really is and touch down without having flared sufficiently to avoid a hard landing.

$$\underline{\qquad\qquad\qquad\qquad\qquad\qquad\qquad\qquad} A272$$

False. Simplifying the explanation, a vortex forms as high-pressure air from beneath the wing curls around the tip in an attempt to fill the "void" of low pressure above the wing.

A273

B. There are two reasons for this. The groundspeed and required heading change are greatest during this turn. This can be confirmed by sketching such a rectangular course and showing the crab direction on each leg.

A274

Many pilots use this expression as a memory jogger for the final items (usually not found on a checklist) to accomplish when cleared for takeoff. "Lights" (turn on the landing lights); "camera" (turn on the transponder so that the aircraft can be seen by ATC and TCAS-equipped aircraft); and "action" (has anyone forgotten to advance the throttle for takeoff?).

A275

D. Climb gradient cannot be determined without knowing groundspeed. The correct answer would be A if the wind were calm.

A276

False. There would not be a warning. An amphibian does not have the intelligence to recognize the type of surface on which it is about to land. The green lights, however, will be illuminated, a case of green lights indicating an unsafe condition.

A277

Wingover, which is roughly similar to a portion of a lazy eight. The maximum bank angle achieved during the maneuver is typically between 45 and 90 degrees.

A278

In general, normally aspirated engines should be leaned and those with turbocharged engines should not. (The pilot's operating handbook, however, is the final authority regarding this subject.)

Engine exhaust is used to heat induction air to melt or prevent the formation of carburetor ice. An idling engine does not develop as much exhaust heat as does an engine delivering power. This is one reason to periodically apply power during a prolonged glide.

B. Similarly, pilots tend to descend prematurely when the lights get brighter or when visibility is unusually good. Bright lights make an object appear closer than it really is.

True. Anyone who has done a belly flop from a diving board can attest to the hardness of water. Small taildraggers equipped with tundra tires can be supported by the "hard" surface of a lake or river at "taxi" speeds of as low as 30 mph or less. (Tundra tires are unusually wide and have relatively low inflation pressures.)

Please do not attempt this without first obtaining the necessary dual instruction. (A281)

_______________________________ A282

The airspeed indicator at such a time is typically unreadable due
to a combination of extremely erratic instrument fluctuations and
involuntary movement of a pilot's eyes within their sockets.

_______________________________ A283

B. The opposite control inputs (control wheel left and fully
   forward) would mean that the wind is coming from behind
   and from the right of the aircraft, which—when heading
   northwest—signifies a wind from the east.

_______________________________ A284

C. Extending wing flaps reduces vortex severity.

_______________________________ A285

False. Because wake turbulence sinks beneath the offending
aircraft, a following aircraft generally encounters a smooth ride
(as long as both aircraft are on the glide slope). A rough ride
can be expected, however, if the flight path of the following
aircraft sinks below that of the lead aircraft.

_______________________________ A286

Not much. The outside loop is begun by pushing the stick forward at the top of the loop and while right side up; the aircraft
becomes inverted at the bottom of the loop. The inverted outside
loop is begun by pushing the stick forward at the bottom of the
loop while inverted; the aircraft becomes right side up at the top
of the loop.

_______________________________ A287

C. The other maneuvers are used to reverse heading. They also
   are called reversements (from the French word, _renversement_).

C. The others are types of stalls. The hammerhead turn is occasionally and incorrectly called a hammerhead stall, but the wings do not stall during this aerobatic maneuver (even though the British call it a stall turn).

A small, anti-spin, drag chute is installed on the tail (usually). Deployment of the chute during a spin arrests rotation and raises the tail (reduces the wings' angle of attack), which facilitates recovery (especially from flat spins). The chute is jettisoned from the aircraft immediately following spin recovery.

The *lomcovák* is a bewildering, seemingly out-of-control maneuver that was developed by the Czechoslovakian and ex-world aerobatic champion, Ladislav Bezak. The airplane flies backwards and sideways, tumbles in somersault fashion, and rotates about all three axes with no apparent forward airspeed.

A great landing is one after which the airplane is still usable.

Turn off the landing lights.

D. An inverted compass card does not rest on its jeweled pivot point as it does during normal flight. It instead is contacting some part of the internal structure of the instrument, which makes card movement erratic and unpredictable.

A294

He is taxiing a seaplane into the wind and toward a docking, beaching, or mooring area and needs to reduce taxi speed as much as possible by increasing drag and decreasing idle thrust. (Seaplanes obviously do not have brakes.)

A295

D. The others are possible characteristics of a stall (in addition to buffeting and activation of a stall-warning system).

A296

False. "Zooming" or "roller-coastering" is inefficient and wastes energy. It is more efficient to maintain altitude until the aircraft has decelerated to its best glide speed.

A297

V_{MU}, or the minimum-unstick speed. V_{MU} can be determined in a lightplane using a procedure similar to that of a soft-field takeoff.

A298

At low airspeeds, the turbulent airflow of reversed air can reenter the engine inlet, which can cause a compressor stall and engine damage. Depending on engine location, low-speed reversing also increases the likelihood of ingesting foreign objects on the runway. Further, thrust reversing loses effectiveness at lower speeds.

A299

E. Although sink rate decreases with a decrease in altitude, true airspeed reduces proportionately, which results in a constant glide ratio.

$$\text{A}300$$

True. Ice covered with water is more slippery and offers less frictional resistance than dry ice. Ice melting under the blades of ice skates allows skaters to achieve greater speeds than if such frictional melting did not occur.

$$\text{A}301$$

The rule of thumb states that a pilot should abort the takeoff if 70 percent of the liftoff speed is not achieved at the halfway point, because it might not be possible to lift off in the remaining distance. Consult the AIM for other considerations.

$$\text{A}302$$

False. The first step is to apply antispin rudder and the second is to reduce angle of attack. This maximizes rudder effectiveness in arresting autorotation.

$$\text{A}303$$

A. When on a northerly heading, northerly turning error initially causes a compass to turn in the wrong direction. When on a southerly heading, it initially causes a compass to move excessively but in the proper direction.

$$\text{A}304$$

An inverted normal loop is begun by pulling the nose down from an inverted attitude; an inverted outside loop is begun by pushing the nose up from an inverted attitude.

$$\text{A}305$$

Don't do anything. With the wings level and in smooth air, an airplane will automatically seek and maintain its trimmed airspeed and angle of attack.

A306

D. This usually results in minimal aircraft damage and no bodily injury. Softer surfaces can result in a part of the aircraft "digging in," and foamed (slippery) surfaces can cause a loss of directional control. These conditions increase the potential for damage and injury.

A307

Oil tends to collect in the bottom cylinders of radial engines. Attempting to start an engine at such a time can result in hydraulically locked cylinders that can cause structural damage to the engine. If a hydraulic lock is detected, the bottom spark plugs are removed and the oil allowed to drain.

A308

False. The tab should be up on the right aileron and down on the left. Although such "left" trim requires more muscle to hold the wheel to the right, the tabs work with, and not against, the ailerons, which makes them more effective at slow speed.

A309

The airplane gets through the downdraft sooner. Also, lowering the nose and reducing power reduces the likelihood of engine overheating that could result from slowing and adding power to maintain altitude. (Lost altitude can be recaptured in the subsequent updraft.)

CHAPTER 3
NAVIGATION

True or False? A minute of latitude over southeastern New Mexico is larger than a minute of longitude in the same area.

Q 311

Why is the time in Greenwich, England (UTC time) referred to as "Z time" or "Zulu time?"

Q 312

True or False? There are 25 time zones covering the Earth.

Q 313

What is the distance between the Equator and the True North Pole? This can be calculated mentally (without pencil and paper).

Q 314

A pilot is cruising at 5,500 feet at a point exactly halfway between the true and magnetic north poles while maintaining a true heading of 045 degrees. The magnetic variation is approximately

    A. 000 degrees.
    B. 090 degrees.
    C. 180 degrees.
    D. None of the above.

Q 315

Suspended cables and wires are not shown on sectional charts unless they are at least

    A. 100 feet AGL.
    B. 200 feet AGL.
    C. 400 feet AGL.
    D. 500 feet AGL.

A pilot maintains a true heading of 045 degrees while flying over the United States in no-wind conditions. The resultant track is a

A. great circle route.
B. small circle route.
C. rhumb line.
D. plumb line.

A pilot using a sectional chart sees a topographical symbol that he does not understand. Where is the only place where all such symbols can be found?

A. the *Aeronautical Information Manual.*
B. the legend on the chart.
C. *Pilot's Handbook of Aeronautical Knowledge.*
D. None of the above.

Which of the following does not belong?

A. magnetic course
B. magnetic heading
C. compass course
D. compass heading

Which of the following states is closest to Africa?

A. Hawaii
B. Maine
C. Florida
D. North Carolina

True or False? While planning a flight, a pilot notes that the wind is forecast to be exactly perpendicular to his true course. This means that groundspeed will be equal to or greater than true airspeed.

True or False? A pilot flying an airplane at a true airspeed of 80 knots is attempting to fly a true course of 360 degrees. The wind is from 270 degrees at 80 knots. It is impossible for the pilot to reach his destination.

Which takes longer —

A round-robin flight with a headwind one way and a tailwind of the same speed while returning,

or

The same round-robin flight except that it is made under the influence of an equally (as above) strong crosswind?

A pilot wants to maintain a true course of 090 degrees. Because he is unaware of a northwesterly wind, he holds a true heading of 090 degrees. Still unaware of the same wind, he reverses course and attempts to fly a true course of 270 degrees using a true heading of 270 degrees. The drift angle in each case is _______________, and the distance that the aircraft will drift off course at the end of one hour in each case is _______________.

    A. the same, the same
    B. the same, different
    C. different, different
    D. different, the same

True or False? A pilot heading toward a DME ground station is within reception range but does not get a DME indication. One possible reason is that too many pilots are attempting to use the station.

A pilot encounters a direct headwind while flying from A to B, points that are 25 *miles* apart. As he passes over B, he releases a balloon that begins to drift back toward A. In the meantime, the pilot continues on his original heading for 15 *minutes*. He then reverses course and returns to A, arriving there at the same time as the balloon. Assuming that no time is required for the airplane to turn, what is the speed of the wind? (A college professor needed three hours to solve this logic problem, but an eighth-grader took only a few minutes using basic arithmetic.)

True or False? A pilot wants to navigate along a military training route designated on his sectional chart as IR3453 at an altitude of 2,500 feet AGL. He will be safely above all military aircraft using that route.

Which states are the most northerly, easterly, southerly, and westerly?

A pilot wants to be at least 1,000 feet above any manmade obstacle no matter where in the 48 states he might be. To do so, he must maintain at least what minimum altitude *above the ground* (not MSL)?

    A. 5,000 feet AGL
    B. 10,000 feet AGL
    C. 15,000 feet AGL
    D. 20,000 feet AGL

A pilot is flying west over Philadelphia during a midday flight, and the sun is above and to his left. What approximate ground-speed must be maintained to keep the sun in the same relative position for the duration of the flight?

A. 690 knots
B. 810 knots
C. 900 knots
D. 1,000 knots

A pilot takes off from the Detroit City Airport in Michigan and heads due south on a long-distance flight. What is the first foreign country over which he will fly?

True or False? A pilot is crabbing left to stay on course while flying in cloud. He is maintaining a constant indicated altitude but is unable to get a current altimeter setting as he proceeds. This means that the pilot is slowly and unwittingly losing altitude.

How is it possible in the *Northern* Hemisphere for a pilot to fly 500 miles north, 500 miles east, and 500 miles south (in that order) so as to arrive at the same point from which he started? (All directions are true and the wind is calm.) Remember, this takes place in the *Northern* Hemisphere.

A pilot wants to fly as far north from his home airport as possible and then return without landing for fuel. Everything else being equal, the greatest radius of action with a given wind speed is achieved when the wind direction is

A. 180 degrees.
B. 210 degrees.
C. 240 degrees.
D. 270 degrees.

Two pilots are flying over tropical islands located on opposite sides of the world while flying identical airplanes that have an unlimited supply of fuel (assume this is possible). Each maintains a true heading of 070 degrees in no-wind conditions until arriving at a common rendezvous point. What is their destination?

True or False? As a pilot proceeds north, meridians of longitude converge.

True or False? Assume that it is noon in Tokyo and 7 p.m. in Los Angeles. The time difference between the two cities, therefore, is seven hours.

Radio aids to navigation in the United States are identified using letters of the alphabet. Which of the following letters are not used to begin the identification of any such NAVAID?

A. Q and W
B. K and Z
C. Q and Z
D. K and W

A pilot is crossing the agonic line (zero magnetic variation) in the vicinity of St. Louis at a latitude of 39 degrees north on a compass heading of 265 degrees. Compass deviation on all headings is -5 degrees. If he maintains this heading under no-wind conditions until reaching the Pacific Ocean, will his latitude increase, decrease, or remain the same, or can this problem not be solved with the information provided?

True or False? Large-scale aeronautical charts provide more detail than small-scale charts.

A pilot is on a nonstop flight under no-wind conditions from Chicago (42 degrees N, 88 degrees W) to Paris, France (49 degrees N, 2 degrees E), via the North Pole. He is approaching the North Pole on a true heading of 360 degrees as shown on his gyroscopic heading indicator. Upon reaching the pole and without resetting the heading indicator, to what indicated heading must he turn so as to track directly toward Paris along a true course of 180 degrees?

A pilot over the 48 conterminous states flies along a parallel of latitude until he reaches his destination, which is 10 degrees of longitude east of his departure point. If the time of sunset at the departure airport is 0020 UTC, what is the time of sunset at the destination?

True or False? Los Angeles, California is west of Reno, Nevada.

Where on Earth does a given day first begin?
   A. immediately east of the International Date Line.
   B. immediately west of the International Date Line.
   C. immediately east of the Greenwich Meridian.
   D. immediately west of the Greenwich Meridian.

.–– .... . .–. . .. .–. . –. ––– ?

    A. .. ––– .–– .–
    B. .. –.. .– .... –––
    C. –. . ...– .– –.. .–
    D. –– .– .. –. .

What is the name of the VOR serving Chattanooga, Tennessee? (Use your imagination.)

On the first day of spring (the vernal equinox), a pilot determines that the sun is 35 degrees above the horizon at local noon (the sun is due south). What is his latitude?

A pilot is flying in cloud under the influence of a strong crosswind. His conventional compass is inoperative and the gyroscopic heading indicator has been set incorrectly. The GPS provides position, track, and ground speed, but not heading or drift angle. How can the heading indicator be reset to the proper heading?

What is the difference between speed and velocity?

True or False? HILLE and BILLI (Clinton) are consecutive intersections on Victor 69 east of Little Rock, Arkansas.

How can a pilot tell the difference between the aural Morse code identifier of a DME transmitter and the same Morse code identifier of the collocated VOR transmitter?

Pilots know that a VOR is a "VHF omni-directional range," but what is a range?

A pilot in an airplane with a true airspeed of 200 knots flies from A to B under the influence of a 50-knot tailwind. The same wind is encountered on the return leg from B to A. Trouble is, the distance between A and B is unknown. Assuming an instantaneous turnaround over B, what is the average groundspeed for the flight?

Underlining the frequency of a VOR station in the frequency box on a sectional chart is used to indicate that communications is not available on that frequency. What is indicated when the name of the station is underlined?

Which of the following states still maintains more than a dozen airway beacons for providing navigational guidance to VFR pilots at night?

A. California
B. Idaho
C. Montana
D. Washington

America's tallest man-made obstruction (in feet above ground level and excluding tethered balloons) is in

A. Florida.
B. Illinois.
C. New York.
D. North Dakota.

Mountains on all sectional aeronautical charts are shaded on one side to depict how they might appear from the air. This shaded relief makes it appear as though the sun is positioned in the

A. northeast.
B. southeast.
C. southwest.
D. northwest.

A symbol on sectional charts is a white dot contained within some symbols for airports with hard-surface runways. This represents the approximate location on the airport of

A. the rotating beacon.
B. the transient tie-down area.
C. customs and immigration.
D. airport police or security, when available.
E. None of the above.

True or False? The 50 states are wholly contained within 10 time zones.

A pilot tunes to a nearby VOR, selects 360 degrees on his OBS (omni-bearing selector) and turns in a southerly direction toward the station. While maintaining a magnetic heading of 165 degrees, the CDI (course deviation indicator) remains halfway between center and a full right deflection. What is the airplane's magnetic track and drift angle?

What official state aeronautical chart currently contains a symbol for and displays the location of a UFO crash site?

Explain how this is possible: The TO-FROM flag is in the OFF position, and the CDI is deflected fully right for 30 minutes. Shortly thereafter and while the aircraft maintains the same track, the CDI swings fully left and remains there while the TO-FROM flag remains in the OFF position.

A wind triangle consists of six elements: true course, true heading, true airspeed, groundspeed, wind speed, and wind direction. Given any four of these elements, it is possible to construct a wind triangle to find the other two. If GS = 142 knots, TC = 037 degrees, TH = 028 degrees, and WD = 333 degrees, determine true airspeed and wind speed.

An airway segment shown as a straight line on a sectional chart connects two VORs. It consists of the 085-degree radial of one VOR and the 270-degree radial of the other. Give three reasons why these radials are not exact reciprocals of each other?

What type of lighted landing facility is associated with each of the following types of rotating beacons?

    A. alternating white and yellow
    B. green, yellow, and white
    C. two white flashes followed by green
    D. alternating white and green

Arrange the major airports at Memphis, Minneapolis, and St. Louis in order of elevation (highest airport first, lowest airport last).

A square inch on a world aeronautical chart represents _______ times as many square miles as a square inch on a terminal area chart.

When the date, December 28, 2010, first occurs on Earth, how much time will elapse before this same date no longer exists anywhere on Earth? In other words, for how long a period of time (in hours) does this or any given date exist somewhere in the world?

Why did the military develop TACAN (tactical air navigation) when a similar navigational system, VOR/DME, was already in civilian use and proven reliable?

When the cartoon character, Tweety, sees Sylvester, a cat, he says, "I tought I taw a puddy tat." How can this expression possibly be of interest to a pilot?

A VFR pilot lifts off from Runway 24, which has a magnetic direction of 245 degrees, under the influence of a crosswind that causes a 5-degree left drift angle. He should climb on an initial magnetic heading of _______ degrees.

The localizer of an instrument landing system (ILS) has a width of

    A. 3 degrees.
    B. 4 degrees.
    C. 5 degrees.
    D. None of the above.

What is the only state in the United States that does not have a border consisting of or containing either a straight line or an arc?

Only two states share common borders with eight other states, which is the greatest number of borders shared by any state. These states are

A. Arkansas and Kentucky.
B. Kentucky and Missouri.
C. Missouri and Tennessee.
D. Tennessee and Arkansas.

What is the world's largest desert?

A. North American Desert
B. Kalahari Desert
C. Sahara Desert
D. None of the above.

Why is the abbreviation for coordinated universal time "UTC" instead of "CUT?"

While flying in the vicinity of an airport, a pilot flies over a marker beacon that transmits a continuous audio identification of two short dots followed by a pause. The white marker-beacon light flashes similarly. What does this signify?

What is the significance of a continental divide?

Two identical, 500-knot airplanes are flying along the Equator in opposite directions at 35,000 feet. (They have already passed each other, so there is no danger of a midair collision.) Assuming that the winds aloft are calm and that each aircraft is identically loaded with payload and fuel, which of the two aircraft has the greatest range?

A pilot is tracking along a highway that has a magnetic course of 355 degrees while maintaining a magnetic heading of 003 degrees and a true airspeed of 150 knots. Without using a computer or calculator, determine the crosswind component.

True or False? Excluding the effects of magnetic deviation, the magnetic compass in an airplane points to the magnetic north pole.

Solar eclipses prove that the sun and moon each occupy (subtend) the same angle of the sky. What is the size of the angle occupied in the sky by the sun or the moon?

A. 0.5 degrees
B. 1.0 degrees
C. 2.0 degrees
D. 4.0 degrees

A pilot tunes in a VORTAC. While attempting to aurally identify the station, she hears a dash, four dots, and a dash that are clustered together as if representing one letter of the alphabet in Morse code. What does this signify?

A pilot of a Cessna 172 is heading directly into the sun on a true heading of 096 degrees and notes that the sun is 15 degrees above the horizon. The pilot of a Piper Warrior is heading into the sun on a true heading of 088 degrees and notes that the sun is 10 degrees above the horizon. The Cessna is __________ of the Piper.

A. southwest
B. southeast
C. northwest
D. northeast

A VOR fix is established by determining the point at which two radials (straight lines) intersect. A DME fix is established by determining the point at which two or three circles intersect. A LORAN fix is determined by the intersection of

A. circles.
B. hyperbolas.
C. parabolas.
D. straight lines.

The four longest rivers in North America are, alphabetically, the Mississippi, the Missouri, the Rio Grande, and the Yukon. If you were to fly along these rivers, arrange them in order of the length of these flights with the longest first and the shortest last.

A pilot is at an airport located at 34 degrees North, 118 degrees West (in the vicinity of Los Angeles). He is about to depart on a flight that will take him to a diametrically opposed point on the other side of the world. What are the geographical coordinates of his destination?

Considerable information can be obtained from a sectional chart by understanding the symbology associated with the rectangular box containing the frequency and identification of a NAVAID. Match each of the following symbols with its closest meaning.

1. Bold-lined box.
2. NAVAID frequency preceded by an asterisk.
3. Frequency followed by an R.
4. Thin-lined box.
5. Small, solid circle containing a white T.
6. Small, solid square in a corner of the box.
7. Underlined frequency.

A. Continuous broadcast of AIRMETs and SIGMETs.
B. Continuous broadcast of weather and aeronautical data.
C. Does not operate continuously.
D. Emergency frequency received by FSS.
E. Frequency of 122.2 MHz not received by FSS.
F. Pilot should transmit but not listen.
G. Voice communications not available.

Global positioning system (GPS) signals are transmitted from satellite to receiver on frequencies in the _______________ band.

A. HF (3-30 MHz)
B. VHF (30-300 MHz)
C. UHF (300-3,000 MHz)
D. SHF (3,000-30,000 MHz)

The title panel of a sectional or other aeronautical chart shows the standard parallels (of latitude) for that chart (a Lambert conformal conic projection). What is the significance of these parallels?

____ **Q 390**

Assume that it is currently 2359 (local time or UTC) on December 28, 2004. What will be the time a minute later at exactly midnight, 2400 or 0000?

____ **Q 391**

Only two VOR radials are required to establish a VOR fix. Why is a minimum of three satellites required to establish a GPS fix?

____ **Q 392**

True or False? GPS receivers typically provide aircraft position in three dimensions (longitude, latitude, and altitude). When climbing vertically at 100 knots in an aerobatic airplane with such a receiver, indicated groundspeed is 100 knots.

____ **Q 393**

The expression, _true virgins make dull company_, is a mnemonic used to help pilots remember the sequence of navigational calculations involving _true_, _variation_, _magnetic_, _deviation_, and _compass_. What is the mnemonic used to recall the reverse order of these calculations?

____ **Q 394**

True or False? A navigational aid used in the United States prior to the introduction of the VOR was the VRB, or voice radio beacon, which broadcast in voice the bearing of an aircraft from the station.

____ **Q 395**

A pilot is navigating toward a 30-mile-distant VORTAC using VOR navigation and notices that the course deviation indicator (CDI) is fluctuating by as much as 6 degrees. The Off flag is hidden from view. He should

    A. change engine rpm.
    B. recycle the selected VORTAC frequency.
    C. reduce airspeed.
    D. report the phenomenon to ATC or a nearby flight service station.

Q 396

True or False? A true course corrected for magnetic variation always results in a magnetic course.

Q 397

A pilot is on a nonstop flight from Los Angeles to Honolulu at FL180 (altimeter set to 29.92 inches Hg) and does not apply any wind correction (crab) whatsoever. The altitude of that flight level above both the departure and destination airports is 18,500 feet MSL. Irrespective of whatever various winds might be encountered en route, the aircraft will pass _____________ Honolulu.

    A. directly over
    B. south of
    C. north of
    D. Insufficient information is provided.

Q 398

The deck of an aircraft carrier is angled 15 degrees left of the ship's centerline, so that the relative wind created by the ship is always at an angle to the "runway." When the surface wind is northerly and equal to the ship's speed, what must the ship's heading be so that the pilot can land without a crosswind? (A computer is not required.)

Q 399

On what radial of the Mission Bay (California) VORTAC is the Bondo intersection located?

Q 400

A pilot makes a forced landing somewhere in the Northern Hemisphere and is uncertain of his position. He sights along the long edge of his plotter and notes from a plumb bob hanging from the center of the plotter that Polaris, the North Star, is 37 degrees above the horizon. What is the pilot's approximate latitude?

True or False? A pilot holding a no-wind heading determines that his aircraft is drifting 10 degrees right of his intended course. If he turns 10 degrees left, this will result in the aircraft exactly paralleling the desired course.

Assume that *Voyager* flew along a great-circle route and at a constant altitude during its historic nonstop, unrefueled flight around the world. How much would the circular distance flown have increased had Dick Rutan and Jeana Yeager flown 1,000 feet higher?

A. 1 nautical mile
B. 6 nautical miles
C. 10 nautical miles
D. 60 nautical miles

A Boeing 747 cruising at Flight Level 350 has a true heading of 360 degrees and an unknown airspeed. The 100-knot wind at that altitude is from 270 degrees true. To an observer on the ground, the contrail appears to be aligned

A. to the east of north.
B. to the west of north.
C. north.
D. The answer cannot be determined with the information provided.

The agonic line extends from near the magnetic north pole to the northeastern tip of Minnesota and to a point abeam Chicago before exiting the United States near

A. Jacksonville, (eastern) Florida.
B. New Orleans, Louisiana.
C. Pensacola, (western) Florida.
D. Savannah, Georgia.

The sun reaches its maximum northerly position over the Earth when it is directly above the Tropic of Cancer (23.5 degrees north latitude) on the first day of summer (in the Northern Hemisphere). This is south of every point in the 48 states. Why, then, must a pilot look northwest (and not southwest) to view the sunset on at least this day from any point in the 48 states?

An airplane with a fully functioning groundspeed indicator is being flown at a constant altitude in no-wind conditions. Why is the pilot not surprised by a groundspeed indication of 0 knots?

The course deviation indicator of a VOR receiver reaches a maximum deflection when the aircraft becomes _____ degrees offset from a selected course. Also, ILS needles reach maximum deflection when _____ degrees left or right of a localizer and _____ degrees above or below a glide slope.

True or False? The airport data block shown on VFR aeronautical charts depict the longest runway at a particular airport as "40." This means that the runway length is longer than 3,950 feet.

True. A minute of latitude anywhere in the world is larger than a minute of longitude in the same area. The only exception is at the equator where a minute of longitude and a minute of latitude have equal length.

Every time zone in the world has an alphabetic designation, and the Greenwich Meridian lies in Zone Z. The Eastern, Central, Mountain, and Pacific time zones are in the R, S, T, and U zones, respectively.

Author's son, Brian, straddling the brass stripe representing the Greenwich Meridian at the Royal Observatory in Greenwich, England. Brian's right foot is in the Western Hemisphere and his left is in the Eastern Hemisphere. (A311)

———————————————————————— A312

True. There are 23 full-size time zones and two half zones that are separated by the International Date Line.

———————————————————————— A313

Since a nautical mile is equal to a minute of latitude, it follows that each degree of latitude is 60 NM. Also, there are 90 degrees of latitude between the Equator and the North Pole. The distance between them, therefore, is 60 times 90, which is 5,400 NM.

———————————————————————— A314

C. If the aircraft is halfway between the true North Pole and the magnetic north pole, then the poles are in opposite directions from the aircraft. The result is that the compass points in one direction (toward the magnetic pole) while the true pole is in the opposite direction. In other words, magnetic variation is approximately 180 degrees (east or west, take your pick).

———————————————————————— A315

B. This is one reason why low-level flight can be hazardous.

———————————————————————— A316

C. A great circle (other than the Equator) crosses meridians at increasing or decreasing angles, and a rhumb line crosses meridians at a constant angle.

———————————————————————— A317

D. All chart symbols are shown and explained in the *Aeronautical Chart User's Guide*, which is published by the National Aeronautical Charting Office (NACO). Every pilot should have a copy of this important booklet.

A318

C. In effect, there is no such thing as a compass course. It can
   be calculated but has no purpose.

A319

B. If you don't believe it, look at a globe.

A320

False. A direct crosswind requires crabbing into the wind,
which reduces groundspeed.

A321

True. It would be nice if an airplane could tack the way a sail-
boat does, but it cannot. Try solving this problem by plotting a
wind triangle.

A322

It takes longer to make a round trip under the influence of a
headwind/tailwind combination than under the influence of a
crosswind.

A323

D. The distance off course in each case is the same, because the
   crosswind component acts upon the aircraft in each case for
   the same amount of time. When flying into a quartering headwind, however, the drift angle is greater than when heading in
   the opposite direction. Try this on an E6-B computer, and you
   will see why.

$$A324$$

True. The DME unit in an airplane is both a transmitter and a receiver. When too many such transmitters attempt to interrogate a DME ground station, the station lowers its own receiver gain until it "hears" only the number of transmissions it can handle. This means that those pilots farthest from the station might be left without a DME indication.

$$A325$$

The airplane travels for 15 minutes away from the balloon. Because the airplane is moving in the same air mass as the balloon, it must also take 15 minutes to return to the balloon. It can be no other way. The airplane, therefore, returns to the balloon over A in 30 minutes. During this time, the balloon has traveled 25 miles. The speed of the wind must be 50 mph.

$$A326$$

True. A military training route with a four-numeral designation is entirely below 1,500 AGL. If any segment of an MTR is at or above 1,500 feet AGL, it will have a three-numeral designation.

$$A327$$

Alaska, Alaska, Hawaii, and Alaska, respectively. Note: Alaska is most easterly because the Aleutian chain stretches to almost 170 degrees east longitude, which means that it is in the Eastern Hemisphere.

$$A328$$

D. There are several large balloons near the southern border of the United States. They are tethered by cable, identified by strobe lights, and tower up to 15,000 feet MSL. The one northeast of Yuma, Arizona, for example, rises to more than 14,000 feet AGL.

____ A329

A. The required groundspeed depends on latitude and varies
   from 900 knots at the Equator to zero at the pole.

____ A330

Canada.

____ A331

True. The left crab indicates that the aircraft is proceeding toward
a low-pressure area. Lacking a current altimeter setting, true altitude becomes less than indicated altitude. This is a variation of
Buys Ballot's Law: "If you stand with your back to the wind, lower
pressure lies to your left in the Northern Hemisphere."

____ A332

As we proceed north from the equator, parallels (circles) of latitude become smaller. If we go far enough north, we eventually
reach a circle of latitude that has a circumference of only 500
miles (*see* figure at right). The problem is solved by starting 500
miles south of this circle. The pilot then flies due north for 500
miles, turns eastward and flies around the North Pole in a 500-
mile circle. At the end of this circumpolar leg, he turns south and
flies for another 500 miles until arriving over the starting point.
(For those technically oriented, a circumpolar track of 500 miles
occurs at 88 degrees 40.4 minutes north latitude. The flight
begins and ends at a latitude 500 miles south of this parallel, or at
80 degrees 20.4 minutes north.)

____ A333

D. The maximum radius of action (for a given fuel endurance
   and true airspeed) occurs when the average out-and-return
   groundspeed is at a maximum, and this occurs when flying
   under the influence of a direct crosswind.

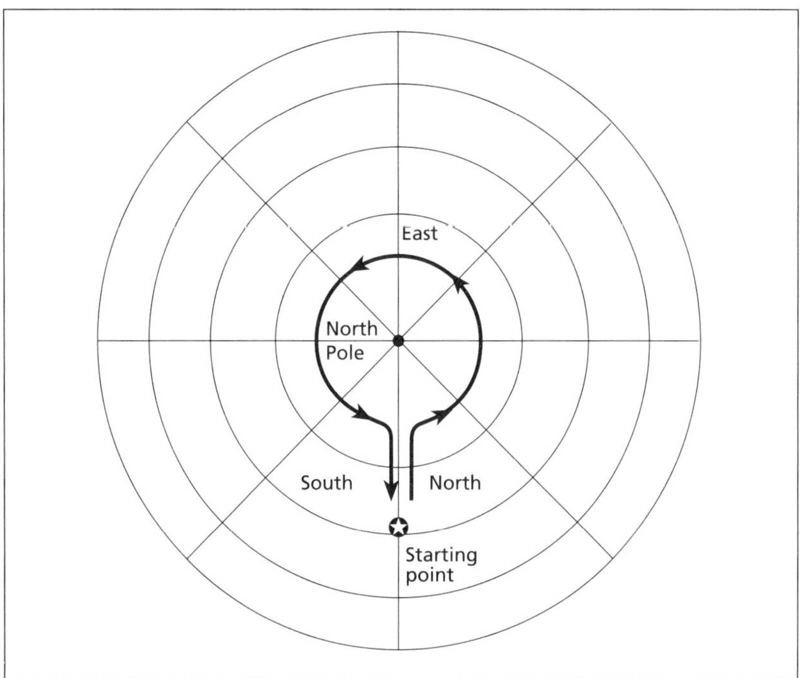

Solution to navigation problem in Question 332. (A332)

—————————————————————————— A334

By maintaining any constant northeasterly (or northwesterly) true heading, each pilot flies a rhumb line that curves around the Earth in an ever-tightening spiral (called a loxodromic curve) about the North Pole, which is where the pilots ultimately arrive after performing ever-tightening, 360-degree turns. Theoretically, they never quite get to the North Pole. But for all intents and purposes, they are there.

—————————————————————————— A335

False. This is true only in the Northern Hemisphere; meridians diverge when proceeding north in the Southern Hemisphere.

False. Because of the International Date Line, the actual time difference is 17 hours.

A. Q and W

The pilot's magnetic heading is 270 degrees. As he proceeds west-bound, however, magnetic variation is easterly and becomes progressively larger. This means that the true heading steadily increases during the flight, which causes latitude to increase.

True. A sectional chart has a larger scale (1:500,000) than a WAC chart (1:1,000,000). With a scale of 1:250,000, a terminal area chart has an even larger scale.

Chicago's meridian of 88 degrees W and Paris's meridian of 2 degrees E intersect at the North Pole at an angle of 90 degrees. The pilot, therefore, must turn 90 degrees to the right, and the new indicated heading must be 090 degrees (even though the airplane is actually heading due south).

2340 UTC. The time of sunset (or sunrise) at the same latitude changes 1 hour every 15 degrees of longitude. (24 hours equals 360 degrees.) Ten degrees, therefore, represents 40 minutes of time. Because an easterly flight shortens the day, sunset occurs 40 minutes earlier.

$$A342$$

False. Reno is 1.34 degrees longitude farther west than Los Angeles.

$$A343$$

B. This means that Guam, which is west of the International Date Line, is the first place in the world to begin celebrating American holidays.

$$A344$$

C. The question is, "Where is Reno?" The answer is "Nevada." The other answers are Iowa, Idaho, and Maine, respectively.

$$A345$$

Choo Choo VOR, which is named after the song, *Chattanooga Choo Choo*, which was popularized by Glenn Miller and Tex Benecke during the Big Band Era.

$$A346$$

55 degrees north. On the first day of spring, the sun is directly over the equator, and the observer's latitude is equal to 90 degrees minus the elevation of the sun. If the pilot were at the North Pole (90 degrees north latitude), the sun would be on the horizon (zero-degrees elevation).

$$A347$$

The pilot enters a shallow turn. When groundspeed (as indicated by the GPS) reaches a minimum or a maximum, the airplane will be aligned with the wind and drift will be zero. The pilot rolls out of the turn and sets his heading indicator to the track shown on the GPS.

$$\overline{\hspace{8cm}} A348$$

Speed describes only the rate of movement; velocity is a vector and describes both speed and direction.

$$\overline{\hspace{8cm}} A349$$

True. Hmm. Perhaps the FAA had a sense of humor during the Clinton administration.

$$\overline{\hspace{8cm}} A350$$

The DME identifier is transmitted at approximately 30-second intervals, only once for every three or four VOR identifiers.

$$\overline{\hspace{8cm}} A351$$

A range is a line of direction, a term borrowed from nautical navigation. The pilot of a ship visually aligns two objects (such as a pair of buoys) and steers along this line (or range).

$$\overline{\hspace{8cm}} A352$$

187.5 knots, irrespective of the distance between A and B. To verify that this is so, determine the out and return times required to fly any arbitrarily selected distance. Then use the total time required to fly the round-trip distance to determine average groundspeed.

$$\overline{\hspace{8cm}} A353$$

The underlined name indicates that the VOR station is a designated visual checkpoint to be used for position reporting.

$$\overline{\hspace{8cm}} A354$$

C. There are 19 such airway beacons in Montana.

D. The KVLY-TV transmitting tower rises to 2,063 feet AGL. It is between Fargo and Grand Forks and can be found on the Twin Cities sectional chart.

D. According to the National Ocean Service, "studies have indicated that our visual perception has been conditioned to this view."

E. The dot shows the approximate location on the airport of a VOR, VORTAC, or VOR-DME.

False. There are only six time zones in the United States: Bering Standard Time, Alaska Standard Time, Pacific Standard Time, Mountain Standard Time, Central Standard Time, and Eastern Standard Time. (Alaska Standard Time is the same as Hawaii Standard Time.)

175 degrees magnetic and 10 degrees right. The aircraft is tracking 5 degrees to the right (or west) of the 360-degree radial, which means that it is tracking inbound to the station along the 355-degree radial, which represents a magnetic track of 175 degrees. (Remember that a half-deflected CDI means that the airplane is 5 degrees off the selected course.) A heading of 165 degrees while tracking 175 degrees indicates that drift is 10 degrees right.

New Mexico. The site is 55 NM from the Chisum (CME)
VORTAC (near Roswell) on the 310-degree radial. It is uncertain
if conventional aircraft are permitted to crash there.

The pilot flies toward, over, and beyond a VOR station with a
selected course that is approximately 90 degrees left of track.

True airspeed = 165 knots and wind speed = 28 knots. The easiest
way to solve this problem is to use an E6-B computer or to construct a wind triangle using pencil, paper, and plotter. Most electronic computers are not programmed to solve such a problem.

Philip Dalton, inventor
of the Dalton E6-B
wind-triangle
computer.   (A362)

A straight line on an aeronautical chart is a great-circle route that changes direction with longitude. Also, magnetic variation can change over a given distance. Also, the reason might be due to misalignment of the VOR antenna array (and 360-degree radial) with respect to magnetic north.

A. seaport
B. heliport
C. military airport
D. civilian airport

Minneapolis (841 feet MSL), St. Louis (605 feet MSL), and Memphis (335 feet MSL). It's an easy question when considering that these airports are adjacent to the Mississippi River, which flows downhill from north to south.

Because each side of a square on a WAC chart represents four times as many miles as one side of an identical square on a TAC chart, the WAC chart contains 16 times the number of square miles.

48 hours. December 28th first occurs immediately west of the International Date Line. Twenty-four hours will elapse before the entire world has the same date, and another 24 hours will be required for the next date, December 29th, to displace the previous date.

____ A368

VOR stations are unsuitable for use on rolling and pitching ships at sea, and they are not adaptable to some of the extraordinary siting conditions required by military operations.

____ A369

The five successive waypoints on the GPS RWY 16 approach to Portsmouth/Pease International Tradeport (KPSM) in New Hampshire are ITAWT, ITAWA, PUDYE, TTATT, and IDEED (see a puddy tat).

____ A370

250. The pilot is expected to track along the extended runway centerline.

____ A371

D. Localizer width depends on runway length and is tailored to be 700 feet wide at the runway threshold. Therefore, the longer the runway, the narrower is the localizer.

____ A372

Hawaii.

____ A373

C. Arkansas and Kentucky share borders with six and seven other states, respectively.

____ A374

D. It is Antarctica, where humidity is typically less than that of the Sahara, and annual snowfall in the interior is less than 2 inches.

A375

An advisory committee of the International Telecommunications Union could not decide in 1970 whether to use the English word order, CUT, or the French word order, TUC, so a compromise, UTC, was adopted.

A376

It is a back-course marker that indicates the location of the final-approach fix of a back-course ILS approach and the point at which the final-approach descent should begin.

A377

It is an extreme stretch of high ground (the Rocky Mountains, for example) from each side of which the river systems of a continent flow in opposite directions.

A378

The eastbound aircraft has the greatest range. This is because its velocity with respect to the center of the Earth is greater than the velocity of the westbound airplane, and the increased centrifugal force as it circles the Earth causes it to weigh 0.4 percent less. This results in less drag and greater range (for a given airspeed).

A379

20 knots. Each degree of crab (or drift) represents a crosswind component of one knot for each 60 knots of airspeed. Therefore, each degree of crab at 150 knots represents 2.5 knots of crosswind. At an airspeed of 60 knots (perhaps when on final approach) each 1 degree of crab represents 1 knot of direct crosswind.

A380

False. A magnetic compass aligns itself with the Earth's lines of magnetic flux, which are influenced by mineral deposits and other magnetic disturbances in the Earth and rarely are aligned with the direction of the magnetic north pole.

____ A381

A. The sun and the moon only appear to be larger. It would
   require 360 full moons (touching at their edges) to span an
   overhead arc of the sky that connects the east and west
   horizons.

____ A382

The station is undergoing maintenance. If the dots and dashes
are mentally and properly separated, the first dash represents a T;
the first dot, an E; the last three dots, an S; and the final dash,
a T, which spells TEST.

____ A383

D. Because both aircraft are heading for the same point, the one
   heading most southerly (the Cessna) must be north of the
   other. Because both pilots are heading east, the one observing
   the higher elevation of the sun (also the Cessna) must be farther east than the other.

____ A384

B. This is why LORAN is known as a hyperbolic navigation
   system.

____ A385

The longest is the Missouri River (2,565 SM), followed by the
Mississippi River (2,350 SM), the Yukon River (1,979 SM), and
finally, the Rio Grande River (1,900 SM).

____ A386

34 degrees south, 62 degrees east (in the Indian Ocean about
2,500 miles east of the southern tip of Africa). The destination latitude must be as far south of the Equator as the departure point is
north of the Equator. The longitude must change by 180 degrees.

1. D. 122.2 MHz also is received.
2. C. May be available on request.
3. F. FSS can receive but not transmit on this frequency.
4. E. The emergency frequency also may be unavailable.
5. B. Transcribed Weather Broadcast (TWEB).
6. A. Hazardous Inflight Weather Advisory Service (HIWAS).
7. G. FSS does not transmit on NAVAID frequency.

C. The 24 satellites (21 plus 3 spares) transmit on two UHF frequencies, 1227.6 and 1575.42 MHz.

Distance and bearing measurements on the chart are most accurate near these parallels. Errors increase slightly as distance from these parallels increases. This is because it is impossible to perfectly portray a spherical surface, the Earth, on a flat sheet of paper. The smaller the scale, the more significant are such errors.

Either is correct as long as it is associated with the correct date. The time could be stated as either 2400 on the day just ending (December 28th) or 0000 on the day just beginning (December 29th).

—————————————————————————————— A391

Each GPS satellite provides a sphere of position (as compared to a line or circle of position). One satellite, therefore, tells a pilot only that his position is somewhere on a sphere, the size of which is determined by his distance from that satellite. Two satellites provide two spheres of position that intersect to form a circle of position. In other words, the aircraft is somewhere on this circle. The sphere defined by the distance from a third satellite intersects this circle of position in two points, either one of which could be the position of the aircraft. Logic circuitry within a GPS receiver eliminates one of these two points as being absurd and selects the remaining point as aircraft position. Without such logic, however, a fourth satellite would be required to resolve the two-point ambiguity. (Technically speaking, because the satellite clocks are not perfect, a solution using three satellites results in an area instead of a point. A fourth satellite is needed to replace the area with a point.)

—————————————————————————————— A392

False. Indicated groundspeed would be zero and is strictly a function of the rate of change of horizontal position. If the airplane were climbing vertically with respect to the air mass, however, wind speed and direction would be shown as groundspeed and track.

—————————————————————————————— A393

There are two: *Can dead men vote twice?* and *Can ducks make vertical turns?*

—————————————————————————————— A394

True. The recorded human voice was broadcast on a rotating beam transmitted on VHF and provided bearings from the station in 10-degree increments.

_______________________________________ A395

A. Certain propeller (or helicopter rotor) speeds (rpm settings)
can cause CDI fluctuations. According to the *Aeronautical
Information Manual*, slight rpm changes will normally smooth
out these fluctuations and should be attempted before reporting unsatisfactory operation.

_______________________________________ A396

True. Deduct an extra point from your score if you anticipated
a trick question.

_______________________________________ A397

A. Whenever flying along a pressure surface (such as FL180)
between two points that have the same altitude (MSL), the
net drift encountered will be nil irrespective of how much
the winds change en route. This is a basic principle of
pressure-pattern navigation (single-heading flight).

_______________________________________ A398

030 degrees. This need not be computed aboard ship, however.
The carrier can simply be turned until the "wind sock" is parallel
to the canted deck.

_______________________________________ A399

The "double-oh-seven" (007) radial, of course.

_______________________________________ A400

37 degrees north. Latitude is equal to the altitude (angular
elevation) of Polaris above the horizon. When the observer is at
the North Pole (90 degrees north latitude), Polaris is directly
overhead (90 degrees above the horizon), and when at the
Equator (zero degrees latitude), Polaris is on the horizon (zero
degrees of elevation).

$$\underline{}\ A401$$

False. The new heading results in the aircraft heading more into the wind, and the new drift angle will be ever so slightly different. For practical purposes, however, drift angle and wind-correction angle are considered to be the same.

$$\underline{}\ A402$$

A. The easiest way to calculate the answer is to recognize that the difference in the circumference of two circles is equal to 2 x π x the difference in their radii, which — in this case — is 2 x 3.14 x 1,000 feet = 6,280 feet. (A nautical mile is 6,080 feet.)

$$\underline{}\ A403$$

C. The airplane and its contrail move together as a single unit within the westerly mass (or flow). The contrail will have the same direction as the heading of the airplane.

$$\underline{}\ A404$$

C. The agonic line connects points of zero magnetic variation and divides the United States into areas of easterly and westerly magnetic variation.

$$\underline{}\ A405$$

When viewing the setting sun, we look in a direction along a great-circle route toward that point on the Earth's surface over which the sun is directly located. Such a great-circle "routing" makes it appear as though the sun is north of our position, even though it is not.

$$\underline{}\ A406$$

The aircraft is equipped with a DME-based groundspeed indicator, and the pilot is flying a circular orbit around a VORTAC. Indicated groundspeed is zero because distance to the station remains constant.

10 degrees (typically), 2.5 degrees, and 0.7 degrees. A glide slope is 3.6 times as sensitive as a localizer and 14.3 times as sensitive as when tracking a VOR radial (for a given distance from the transmitters).

False. The rounding-up point for runway lengths on VFR aeronautical charts is 70, not 50 feet. This means that the actual runway length is between 3,970 and 4,069 feet.

CHAPTER 4
WEATHER

Which is colder, -40° Celsius or -40° Fahrenheit?

_____________________________ **Q 410**

Which of the following does not belong?

    A. radiation
    B. upslope
    C. downslope
    D. advection

_____________________________ **Q 411**

The number of thunderstorms that occur daily over the surface of the Earth is approximately

    A. 1,000.
    B. 5,000.
    C. 10,000.
    D. 50,000.

_____________________________ **Q 412**

During standard conditions, at which altitude will atmospheric pressure be half of what it normally is at sea level?

    A. 15,000 feet
    B. 17,000 feet
    C. 18,000 feet
    D. 20,000 feet

_____________________________ **Q 413**

Which of the following does not belong?

    A. adiabatic
    B. orographic
    C. frontal
    D. air mass

_____________________________ **Q 414**

True or False? Water vapor is lighter than air.

A parcel of air has a temperature of 80° Fahrenheit and a relative humidity of 50 percent. If its temperature is raised to 100 degrees and no water vapor is added or taken away, its relative humidity will be

A. 100 percent.
B. 75 percent.
C. 50 percent.
D. 25 percent.

Which of the following does not belong?

A. Heavy rain from a thunderstorm
B. Virga
C. Trees bending with the wind
D. A ring of blowing dust

When ice on a wing becomes water vapor (a gas) without going through the liquid (water) stage, this is called ________. The reverse process (when water vapor becomes snow or ice without passing through the liquid phase) is called ________.

True or False? In the Southern Hemisphere, wind flows counter-clockwise about a high-pressure area, clockwise about a low-pressure area, and from a low to a high.

Structural icing can occur with the outside air temperature as high as +5° Celsius, which is well above the freezing point of water. How is this possible?

Questions

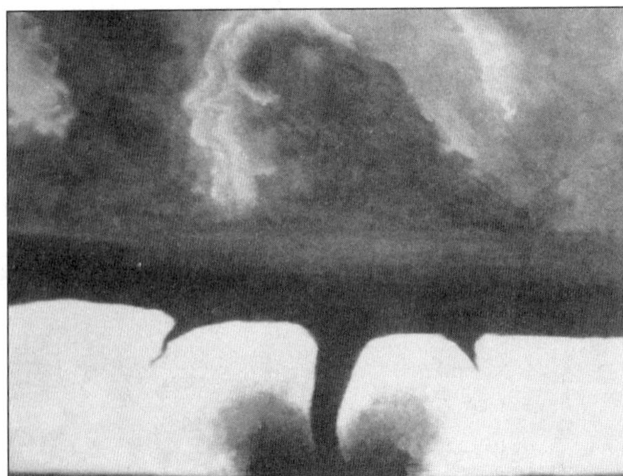

Oldest known photograph of a tornado, dated 1884. (Courtesy of N.O.A.A.) (Q420)

Q 420

What is the difference between a tornado and a funnel cloud?

Q 421

The horse latitudes circle the earth at about 30 degrees north and south latitude and are characterized by light surface winds. Why are they so named?

Q 422

A pilot can obtain a rough estimate of the movement of a thunderstorm by referring to the wind velocity at

    A. 2,000 feet.
    B. 5,000 feet.
    C. 10,000 feet.
    D. 18,000 feet.

Q 423

True or False? A pilot is flying a jet airplane at a true airspeed of 460 knots through a cloud containing supercooled water droplets. The ambient (static) air temperature is -5°C. He should anticipate structural icing.

True or False? Mid-latitude winds aloft in the Northern Hemisphere consist of prevailing westerlies, which is why the prevailing wind over the United States blows from west to east. Mid-latitude winds aloft in the Southern Hemisphere — such as over Australia — also consist of prevailing westerlies.

A pilot is about to depart an airport that has an elevation of 6,000 feet MSL. The altimeter setting is 29.92 inches Hg and humidity is nil. Above what minimum reported temperature would the density altitude be greater than the elevation?

A cold front overtakes a warm front and lifts off the ground the relatively warm air mass that had separated the fronts. What is this phenomenon called?

The frequency of lightning flashes generated by a cumulonimbus cloud is an approximate measure of the thunderstorm's

   A. diameter.
   B. height.
   C. potential for structural icing.
   D. turbulence.
   E. precipitation.

What do the words *scirocco*, *chinook*, *mistral*, and *bora* have in common?

Which of the following does not belong?
    A. convection
    B. ocean surface
    C. surface irregularities
    D. atmospheric gravity (lee) waves
    E. wind shear

Many pilots use the terms chop and turbulence interchangeably, but they are not the same. What is the difference between them?

Which of the following does not belong?
    A. winds shift to northwesterly
    B. temperature decreases
    C. dew point increases
    D. pressure rises

True or False? Thunderstorms often penetrate the tropopause and extend thousands of feet into the stratosphere.

According to FAA's publication, *Aviation Weather*, there are six basic types of precipitation. Can you name them?

A pilot preflights his airplane after a clear night and just before sunrise. There is frost on the wings but no ice or moisture on the ground. How is this possible?

Q 435

The wind over the mid-latitudes is generally westerly. What is the general wind direction over the high and low latitudes in the Northern Hemisphere?

Q 436

A cube of air measuring 1,000 feet per side (the bottom of which is at sea level) has a mass of approximately

    A. 750 pounds.
    B. 7,500 pounds.
    C. 750,000 pounds.
    D. 75 million pounds.

Q 437

If the dew point is 87° Fahrenheit, the increase in density altitude resulting from humidity alone is equivalent to increasing the temperature by

    A. 0°.
    B. 4°.
    C. 9°.
    D. Cannot be determined.

Q 438

A pilot boarding his airplane sees a bolt of lightning in the distance. The associated clap of thunder is heard 18 seconds later. The distance to the thunderstorm is approximately

    A. 2.3 miles.
    B. 3.0 miles.
    C. 3.6 miles.
    D. 4.5 miles.

Q 439

What causes fair-weather cumulus clouds to form over oceans and large bodies of water?

For what purpose have meteorologists found hair from a blonde female to be superior to that of any other person?

Describe a relatively easy method of converting degrees Celsius into degrees Fahrenheit without using a chart, calculator, or pen and paper.

On an airport surface observation, "zero-zero" conditions used to be designated by the alphanumeric coding, "W0X0F" (indefinite ceiling, zero; sky obscured; visibility, zero; fog). How are the same conditions designated in a METAR report?

Fair-weather cumulus clouds are forming overhead at a time when the surface temperature is 80°F and the dew point is 58°F. The base of the clouds, therefore, is

A. 5,000 feet AGL.
B. 6,000 feet AGL.
C. 7,000 feet AGL.
D. 8,000 feet AGL.

With respect to flight in icing conditions, when the outside air temperature is between 0° and -10°C, _______ type of structural ice can generally be expected; when the OAT is between -15° and -20°C, _______ type of ice can be anticipated.

In general, why is there more cloudiness associated with a low-pressure system than a high?

For atmospheric conditions to be unstable, the adiabatic lapse rate generally must be _________ the ambient lapse rate.

   A. greater than
   B. equal to
   C. less than
   D. Does not matter.

If the air at a given location is warmer than standard, the pressure lapse rate (everything else being equal) at sea level is

   A. more than 1 inch of Hg per 1,000 feet.
   B. less than 1 inch of Hg per 1,000 feet.
   C. 1 inch per 1,000 feet.
   D. Cannot be determined.

According to FAA's publication, *Aviation Weather*, the minimum wind speed of a jet stream is _________ knots, and the minimum sustained wind speed within a hurricane (typhoon) is _________ knots.

Can you estimate within 10 knots the highest surface wind ever recorded anywhere (and not associated with a tornado)?

What is the international (ICAO) definition of a broken layer?

The scientific name for the northern lights is *aurora borealis*. What is the scientific name for the southern lights, which occur in the Southern Hemisphere?

True or False? The United States has more tornadoes each year than any other country, and it also has the strongest tornadoes.

Strong, dry, warm winds that flow down the lee slopes of North American mountain ranges are called _______.

True or False? In the Northern Hemisphere, Coriolis force causes fluids to circulate clockwise about high pressure and counterclockwise about low. Therefore, Coriolis force causes the water in a flushed toilet bowl in the United States to drain counterclockwise.

True or False? The word, *virga*, which means rain or ice particles that evaporate before reaching the ground, is an acronym.

Pilots know the general meaning of scud running, but what is the meteorological definition of scud?

Which of the following cities annually experiences more lightning than any other in the United States?

    A. Albuquerque, New Mexico
    B. Atlanta, Georgia
    C. Kansas City, Missouri
    D. Tampa, Florida

Most pilots know that 0° Celsius represents the freezing point of water. What is the significance of 0° Fahrenheit?

--- **Q 459**

With respect to the METAR weather format, what is the abbreviation for sleet?

--- **Q 460**

What is the difference between severe and extreme turbulence?

--- **Q 461**

A pilot can encounter structural icing when flying through a cloud consisting of supercooled water droplets (water that remains in a liquid state below zero degrees Celsius). Why do these droplets not freeze when the ambient temperature is below freezing?

--- **Q 462**

St. Elmo's fire is a luminous discharge of static electricity that appears on (and sometimes in) an aircraft during flight in strong electrical fields, especially when in clouds in proximity to thunderstorm activity. Who is St. Elmo?

--- **Q 463**

True or False? There is a place on Earth where — according to weather records — there has never been so much as a trace of rain or snow.

--- **Q 464**

Which of the following does not belong?
  A. cumulonimbus
  B. altostratus
  C. mammatocumulus
  D. nimbostratus

--- **Q 465**

Why is that sultry time of year between early July and early September referred to as the "dog days of summer?"

Which of the following does not belong?

    A. wind shift
    B. pressure change
    C. temperature change
    D. precipitation change

A brave helicopter pilot takes off and climbs vertically as the surface warm front of a warm front occlusion approaches his ground position. As he climbs, he will first be in the

    A. warm air mass, then the cold air mass, and then the cool air mass.
    B. cool air mass, then the warm air mass, and then the cold air mass.
    C. cold air mass, then the cool air mass, and then the warm air mass.
    D. cool air mass, then the cold air mass, and then the warm air mass.

True or False? If frontogenesis describes the birth of a front, then frontolysis is a term used to describe the dissipation of a front.

What insect can be used to determine temperature to within 1° Fahrenheit, and how is this done?

A pilot encounters a visibility restriction due to "VOG." In what state is he most likely flying?

The expression "he or she is on cloud nine" describes a feeling of euphoric elation and joy. What is the meteorological origin of this expression?

True or False? If half of the sky is obscured by cloud at a given height above the ground, this is considered to be a broken (BKN) layer.

True or False? A batted baseball travels farther in humid air than in dry air.

The absence of a ceiling (or sky condition) and visibility on an ATIS broadcast indicates that these have been observed to be at least

    A. 3,000 feet and 3 miles.
    B. 3,000 feet and 5 miles.
    C. 5,000 feet and 5 miles.
    D. 5,000 feet and 8 miles.

True or False? An old saw cautions not to eat yellow snow, but blankets of yellow snow really do occur in nature.

At 9,927 MSL, Leadville, Colorado, is the most highly elevated airport in the United States. What is the temperature at Leadville when the density altitude there is the same as its elevation (assuming dry air and an altimeter setting of 29.92 inches of Hg)?

True or False? Microbursts are known to have vertical velocities in excess of 1,800 fpm at only 200 feet above the ground.

Contrails (or condensation trails) can form behind jet engines at high altitude when hot, moist exhaust mixes with cold, dry ambient air. What are distrails?

_______________________________________ **Q 479**

During a preflight weather briefing, a pilot notes that "ACSL" is mentioned in the "remarks" section of a METAR report. What do these letters represent and what conditions can be expected?

_______________________________________ **Q 480**

Which state typically experiences the most days of thunderstorm activity during the three winter months?

_______________________________________ **Q 481**

True or False? Hurricanes and typhoons are meteorologically identical.

_______________________________________ **Q 482**

According to the *Aeronautical Information Manual*, the intensity of snowfall can be estimated by the visibility restriction it causes. It is heavy when the visibility is less than ________ statute mile; it is light when more than ________ SM; and it is moderate when visibility is between these values.

   A. 1/8, 1/4
   B. 1/4, 1/2
   C. 1/2, 1.0
   D. 3/4, 1.5

_______________________________________ **Q 483**

True or False? Low-pressure troughs and high-pressure ridges associated with a jet stream in the Northern Hemisphere are situated, respectively, north and south of the jet.

_______________________________________ **Q 484**

When flying over relatively flat terrain, the lowest altitude at which a jet stream can be encountered is

   A. 20,000 feet AGL.
   B. 10,000 feel AGL.
   C. 5,000 feet AGL.
   D. zero feet AGL.

The most prolific breeding ground in the world for tornadoes is in the United States and is called Tornado Alley, which consists primarily of four states. What are the two states (or parts of states) in Tornado Alley that annually spawn the greatest number of tornadoes per unit area?

Estimate within a quarter-inch of mercury the greatest atmospheric pressure ever recorded.

Tornadoes are born in the bellies of thunderstorms and most often are found in the ______________ portion of such storms.

    A. northeast
    B. southeast
    C. southwest
    D. northwest

There are ten main types of clouds. Nine of them are Ci, Cs, Cc, As, Ac, St, Sc, Cb and Cu. What is the missing cloud type? (Note: towering cumulus is not a basic cloud type.)

True or False? Intense snowfall rates can be generated by winter thunderstorms (in New York, for example).

Meteorology is the science that deals with atmospheric phenomena, especially weather and weather conditions. It has nothing whatever to do with meteors, so why is this field of study so named?

Weather in the United States generally is influenced by three types of air masses: cP, mP, and mT. Everything being typical, arrange these air masses in order of their moisture content, most humid first and driest last.

In 1803, Englishman Luke Howard presented the scientific world with the current cloud-classification system, which uses Latin words to describe basic cloud shapes. For example, cumulus means *heap*, and stratus means *layer*. He referred to high, wispy clouds as cirrus, which means _______________________.

True or False? Mountain waves and their associated rotors occur only when the air is stable.

The greatest wind speed ever measured near the surface of the Earth is
  A. 268 mph (233 knots).
  B. 318 mph (277 knots).
  C. 368 mph (320 knots).
  D. 418 mph (363 knots).

A flash of lightning lasts for only an instant. Why does thunder, which is the sound made by lightning, usually last so much longer?

True or False? Hail is one possible danger associated with flight over a large wildfire.

What is the difference between a microburst and a downburst?

Which of the following are certified to fly in moderate freezing rain?

    A. any airplane certified for flight into known icing conditions
    B. large turbofan jetliners
    C. aircraft that cruise in excess of Mach 0.85
    D. any turbofan aircraft using engine bleed air for anti-icing

True or False? It really can rain cats and dogs.

State within 10° Fahrenheit the coldest temperature (exclusive of wind-chill factor) ever recorded on Earth in the free atmosphere.

Reunion, a tiny island in the South Indian Ocean, holds several word records for rainfall. During a 24-hour period beginning on January 7, 1966, the measured rainfall there was

    A. 26.80 inches.
    B. 41.80 inches.
    C. 56.80 inches.
    D. 71.80 inches.

True or False? All snowflakes are six-sided or have six points, and no two snowflakes are alike.

____ **Q 503**

A pilot is preflighting a Cessna 172 in the vicinity of thunderstorm activity when his skin suddenly begins to tingle and his hair stands straight up (à la boxing promoter Don King). What should he do?

____ **Q 504**

A pilot might say that he is in the doldrums when feeling sluggish, gloomy or in a low-energy state in which something fails to develop or improve. What is the meteorological source of this expression?

____ **Q 505**

What is the most practical way for a pilot to protect against hydrometeors?

____ **Q 506**

Why does outside air temperature decrease with a gain in altitude?

____ **Q 507**

A pilot flying close to and above a cloud layer sees the sharply defined shadow of his aircraft completely encircled by a rainbow-like ring. This is called a _____________.

____ **Q 508**

True or False? A strong wind is blowing across a narrow gorge or canyon. The strongest downdrafts are found on the downwind side of the gorge.

____ **Q 509**

Estimate within 5° Fahrenheit the hottest temperature ever recorded in the free atmosphere.

Low-latitude winds in the Northern Hemisphere are northeasterly trade winds. Low-latitude winds in the Southern Hemisphere are

A. northeasterly.
B. southeasterly.
C. southwesterly.
D. northwesterly.

True or False? A pilot holding in position for takeoff on Runway 30L in St. Louis on a fair-weather day notices that some of the low, scattered clouds cast shadows that track from the northeast while the other low clouds are moving toward the south. The lower of the two cloud layers, therefore, is from the north.

The most unfavorable weather associated with a low-pressure system is most often found in the _______________ quadrant.

A. northeast
B. southeast
C. southwest
D. northwest

In middle latitudes, the bases of low clouds (stratus, stratocumulus, nimbostratus, and fair-weather cumulus) are normally found anywhere from the surface up to a height of _______________, and the bases of high clouds (cirrus, cirrostratus, and cirrocumulus) are normally found at a height at or above _____________ feet.

A. 6,500 feet, 14,500 feet
B. 6,500 feet, 16,500 feet
C. 8,500 feet, 14,500 feet
D. 8,500 feet, 16,500 feet

The shortest day of the year (the winter solstice) occurs on or about December 21st and is when the Northern Hemisphere receives the least amount of sunlight. Why is it that February — during which there is more sunlight than in December — is the coldest month?

Which of the following does not belong?

- A. cyclone
- B. hurricane
- C. tornado
- D. typhoon

The Fahrenheit and Centigrade scales cross at -40°, which means that -40°F equals -40°C. Neither is colder than the other. Above -40°, temperature in Centigrade is always warmer than the same number of degrees Fahrenheit, and vice versa.

_________________________________ A410

C. Radiation, upslope, and advection are types of fog. There is no such thing as downslope fog.

_________________________________ A411

D. There have been times when I thought I had to circumnavigate all of them.

_________________________________ A412

C. The atmospheric pressure at 18,000 feet on a standard day is 14.94 inches of mercury, which is almost exactly half of the standard sea-level pressure of 29.92 inches.

_________________________________ A413

A. There are orographic, frontal, and air-mass thunderstorms, but there is no such thing as an adiabatic thunderstorm.

_________________________________ A414

True. This is why humid air — which contains water vapor — is less dense than dry air, which contains no water vapor.

_________________________________ A415

D. For every 20 degrees that temperature increases, relative humidity decreases by 50 percent. Conversely, for every 20-degree decrease, relative humidity doubles.

_____________________________________ A416

C. The other items are visual signs indicating the possible presence of a microburst.

_____________________________________ A417

Sublimation (in both cases).

_____________________________________ A418

False. The first two statements are correct, but wind flows from a high to a low in both hemispheres.

_____________________________________ A419

Air flowing about wings and other parts of the structure increases in velocity (because of venturi effect). This results in local reductions of static pressure, which reduce local air temperature.

_____________________________________ A420

A tornado is a funnel cloud that touches the ground. A funnel cloud that touches a body of water is called a water spout.

_____________________________________ A421

Ancient sailing vessels made poor progress with so little wind to propel them. Horses were thrown overboard to lighten the load.

_____________________________________ A422

C. The speed and direction of the wind at 10,000 feet are roughly the same as the velocity of the thunderstorm.

_____________________________________ A423

False. At 460 knots, the temperature of the air immediately ahead of the aircraft is increased about 30° because of compression, resulting in a total air temperature of 25°, which is too warm for structural icing to occur.

$$\overline{\hspace{7cm}} A_{\!}424$$

True. Not everything in the Southern Hemisphere is backwards and upside down.

$$\overline{\hspace{7cm}} A_{\!}425$$

38°F, which is determined by subtracting 3.5°F for each 1,000 feet of elevation from the standard temperature (at sea level) of 59°F. In other words, 38° Fahrenheit is the standard temperature for an altitude/elevation of 6,000 feet.

$$\overline{\hspace{7cm}} A_{\!}426$$

An occluded front (or an occlusion).

$$\overline{\hspace{7cm}} A_{\!}427$$

D. Lightning frequency provides a general indication of turbulence intensity.

This is not what a pilot wants to see through his windshield. (A427)

___________________________________ A428

Each is the name given to a local wind condition. A scirocco,
for example, is a hot, humid, southerly wind of southern Italy,
Sicily, and the Mediterranean islands.

___________________________________ A429

B. The other choices represent the four basic causes of turb-
    ulence.

___________________________________ A430

Light or moderate chop is a form of light or moderate turbulence that is characterized by rapid and rhythmic bumpiness
that does not appreciably affect attitude, altitude, or airspeed.
Other forms of turbulence produce sporadic jolts.

___________________________________ A431

C. After a cold front passage, dew point decreases.

___________________________________ A432

True. The amount of overshoot into the stratosphere depends
on the strength of the updrafts and the stability of the lower stratosphere.

___________________________________ A433

Drizzle, rain, snow, ice pellets, hail, and ice crystals.

___________________________________ A434

There are several reasons, but the most common is that metal wings
lose (radiate) heat more rapidly than the ground and get colder.

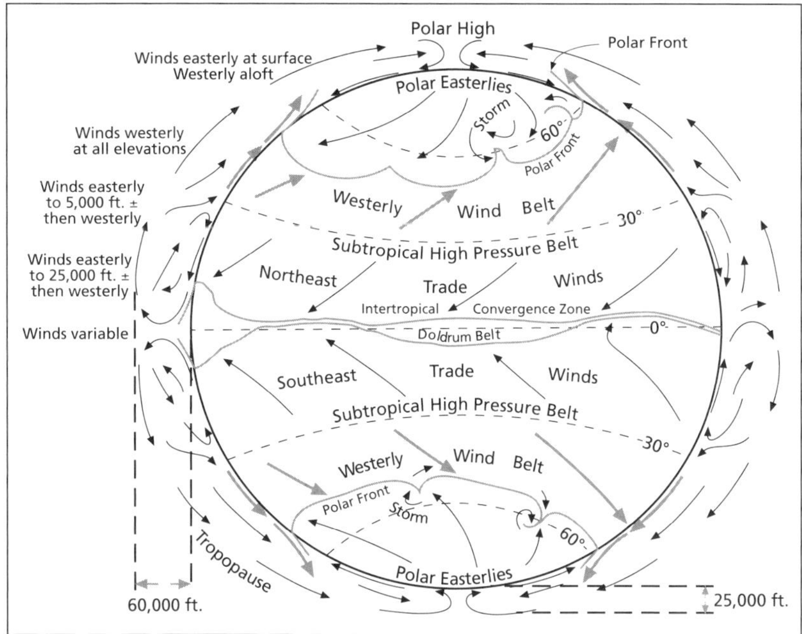

Idealized pattern of atmospheric circulation.   (A435)

$$A435$$

Polar easterlies prevail in the Arctic, and northeasterly trade winds prevail in the tropics.

$$A436$$

D. Such a cube contains 1 billion cubic feet, and each cubic foot has an average density of 0.075 pounds per cubic foot. One billion x 0.075 = 75 million.

$$A437$$

C. High humidity has a greater effect than generally is appreciated.

$$A438$$

C. At sea level, sound travels approximately one mile every five seconds (1,100 feet per second).

(A439)

____ A439

When cool air moves horizontally over relatively warm water,
it is heated from below and forced to rise. Cumulus clouds form
if the dew point is sufficiently high and if the temperature difference between the air and the water is sufficiently large.

____ A440

Blonde female hair reacts most to temperature and humidity
changes (as most natural blondes can attest). It has been used in
radiosondes carried aloft in balloons to detect upper-air conditions.

____ A441

Double the value of Celsius, deduct 10 percent of the result, and
add 32. For example, if the temperature is 20° Celsius, doubling
that would be 40. Ten percent is 4, which when subtracted from
40 equals 36. Adding 32 results in a temperature of 68° Fahrenheit.

A442

0SM FG VV000 (visibility, zero statute miles; fog; vertical visibility, zero).

A443

A. Cumulus clouds are formed by rising air that cools at the dry adiabatic lapse rate of 5.4°F per 1,000 feet. The dew point decreases 1°F per 1,000 feet. They converge, therefore, at 4.4°F per 1,000 feet. Dividing the dew point–temperature spread (22 degrees) by 4.4 gives the cloud base in 1,000s of feet. (In other words, the cloud base occurs at that altitude where the temperature and the dew point become the same.)

A444

Clear, rime. Mixed clear and rime generally can be expected between -10° and -15°C (according to FAA Advisory Circular 91-51A).

A445

A low is characterized by rising air, which cools adiabatically and is a primary cause of cloudiness. A high is characterized by subsidence (descending air), which results in warming and a decrease in cloudiness.

A446

C. In this way, rising air being cooled adiabatically remains warmer than the surrounding air and continues to rise on its own accord.

A447

B. Because warm air is less dense than cold air, it takes a greater gain in altitude for pressure to decrease a given amount when climbing in warm air than cold.

_______________________________________ A448

50, 65

_______________________________________ A449

A low-level jet stream of 201 knots (231 mph) was recorded at
the summit of Mount Washington, New Hampshire on April 12,
1934 at an elevation of only 6,288 feet MSL.

_______________________________________ A450

The International Congress of Agricultural Organizations defines
a broken layer as a nonproductive hen. (Sorry about that.)

_______________________________________ A451

Aurora australis.

_______________________________________ A452

True. No other place on Earth comes close to matching the
number and severity of tornadoes in the United States.

_______________________________________ A453

Chinooks. Internationally, they are known as foehn winds (especially
in German-speaking countries). One of the best-known chinooks
(or *foehns*) in the United States is southern California's Santa Ana
wind that blows from the high Mojave Desert into the lower Los
Angeles Basin (toward the city of Santa Ana). The wind flows
downhill and becomes warmer because of adiabatic compression.

A454

False. A draining toilet bowl is too small to be affected by Coriolis force. Fluid flow rotates in either direction with equal probability and is determined by other factors. Coriolis effect can be detected by draining a very large lake or swimming pool that has stood still for hours.

A455

False. Virga is derived from a Latin word meaning *spray*. The notion that virga is an acronym derived from *Vermogenig Indizienbeewis Regen Geschehenligh Ahoc* is a myth.

A456

Technically, scud is either fractostratus or fractocumulus of bad weather. It usually consists of small, dark clouds that hang beneath a layer of stratocumulus or nimbostratus.

A457

D. The other cities do not experience winter thunderstorms to the extent that Tampa does.

A458

Zero degrees Fahrenheit is the temperature at which salt no longer prevents water from freezing.

A459

Sleet is essentially a British term for ice pellets, the METAR code for which is PE. Ice pellets (or sleet) forms when melted snow passes through frigid air (usually ahead of a warm front) and refreezes.

193   **Answers**

______________________________________ A460

Severe turbulence causes large, abrupt changes in altitude and/or attitude, and the aircraft may be momentarily out of control. Extreme turbulence makes aircraft control practically impossible and may cause structural damage.

______________________________________ A461

The most common reason is that water vapor condenses on microscopic salt particles (hygroscopic nuclei), which lowers the freezing point of the droplets. When disturbed by an airplane, the droplets can instantly change to ice.

______________________________________ A462

St. Elmo (d. A.D. 303) is the patron saint of sailors, who observed this phenomenon as static electricity dancing on the masts of ancient sailing vessels. Reducing airspeed usually lessens the intensity of St. Elmo's fire.

St. Erasmus, nicknamed "St. Elmo." (Detail from a painting by M. Grünewald, ca. 1520-1524.) (A462)

______________________________________ A463

True. The driest place on Earth is Calama, which is in the Atacama Desert in Chile.

A464

C. Each of the others is one of the principal cloud types; mammatocumulus is not.

A465

This is when Sirius, the Dog Star, rises and sets with the sun.

A466

D. The other conditions indicate a frontal passage. Dry fronts have little or no precipitation associated with them.

A467

C. In any frontal system, the coldest air is at the bottom and the warmest air is at the top.

A468

True. These terms are used in conjunction with warm, cold, and occluded fronts.

A469

Temperature in Fahrenheit can be determined by counting the number of chirps made by an ordinary house cricket (saltatorial orthopterous) in 15 seconds and adding the result to 37.

A470

The contraction VOG (volcanic "fog") is used in Hawaii to describe a visibility restriction due to volcanic smoke in the way that the word, smog, is used in some large cities to describe "smoke and fog."

A 471

Cirrus was once considered to be the highest cloud and was ninth on the list of basic clouds. Someone "on a cirrus cloud" or on "cloud nine" was high indeed. We now know that the tops of some thunderstorms are higher, and someone on that, the tenth cloud, would be high indeed (and in store for a ride that might not be so euphoric).

A 472

False. It is a scattered (SCT) layer because it obscures only 3/8ths to 4/8ths of the sky. A broken (BKN) layer obscures 5/8ths to 7/8ths of the sky.

A 473

True. Humid air contains water vapor, which is less dense than dry air and, therefore, results in less drag, which aids the ball in flight. This is why humidity elevates density altitude.

A 474

C. When reported, ceiling is expressed in feet above ground level (AGL), and visibility is expressed in statute miles.

A 475

True. Yellow snow can occur when substantial quantities of pollen are blown into the atmosphere and serve as hygroscopic nuclei about which snowflakes form.

____ A476

-5° Celsius (or 23° Fahrenheit), which is the standard temperature for an elevation of 9,927 feet MSL. Standard temperature for a given elevation is determined by subtracting 2°C per 1,000 feet of elevation from the standard temperature at sea level of 15°C. It is worth noting as an example that density altitude at Leadville is 13,100 feet when the temperature on the ground there is only 75°F.

____ A477

True. To qualify as a downburst or microburst, vertical velocity must be at least 720 fpm at 300 AGL.

____ A478

Distrails (or dissipation trails) are streaks of clearing that occur behind a airplane as it flies near the top of or barely within a thin cloud layer. The heat of exhaust and/or the mixing of dry downwash air from the aircraft causes the cloud to dissipate.

____ A479

"Altocumulus standing lenticular" clouds indicate the presence of a mountain wave.

____ A480

Louisiana averages at least 10 days of thunderstorm activity during the winter. (Florida is the thunderstorm leader during the summer, when it averages more than 50 days of activity.)

____ A481

True. The only difference between them is their location. Tropical cyclones in the Western Hemisphere are called hurricanes; those west of the International Date Line are called typhoons.

_____________________________________ A482

B. Light, moderate and heavy drizzle intensity (but not rain
   intensity) can be estimated using the same values of visibility.

_____________________________________ A483

True. The jet stream typically meanders snakelike in an easterly
direction and forms the bottom (southerly) arcs of troughs and
the top (northerly) arcs of ridges.

_____________________________________ A484

D. Although rare, jet stream slivers can plunge to the surface
   of the Earth and sandblast the area with high-velocity wind.
   This phenomenon can last from less than a minute to several
   minutes.

_____________________________________ A485

Oklahoma and northern Texas. The other two states forming
Tornado Alley are Kansas and Nebraska.

_____________________________________ A486

32.01 inches Hg at Agata, Siberia on December 31, 1968. The
greatest recorded in the United States was 31.85 inches Hg at
Barrow, Alaska on January 31, 1989.

_____________________________________ A487

C. As a visual aid in determining the safest direction of thunderstorm circumnavigation, consider that tornadoes usually
   descend from a rotating wall cloud that often forms the lowest
   portion of the thunderstorm cloud base.

A488

Ns (nimbostratus), which is a stratus cloud that makes the sky appear uniformly gray and produces precipitation.

A489

True. Called thunder snow, this occurs when violent up- and downdrafts at sub-freezing temperatures discourage the melting of snow and ice into rain.

A490

Ancient Greeks used the word *meteorologik* to refer to any object appearing in the sky, including clouds and shooting stars. Eventually, flaming rocks falling through the atmosphere became a part of astronomy, while weather-related phenomena continue to be classified as meteorological.

A491

mT (maritime tropical), mP (maritime polar), and cP (continental polar). Maritime air masses originate over water, and continental air masses originate over land. Tropical air masses generally contain more moisture than polar air masses.

A492

Cirrus means a *lock of hair*, not *mare's tail*, which often is used to describe the appearance of this high-altitude cloud (above 16,500 feet).

A493

True. The waves are caused by vertically displaced, stable air attempting with ever-decreasing amplitude to return to its natural level. Atmospheric instability would cause wind deflected by mountain slopes to rise unabatedly.

The F-Scale (Fujita Tornado Damage Scale)
Developed in 1971 by T. Theodore Fujita of the University of Chicago.
(For further information, see the Storm Prediction Center website at http://www.
spc.noaa.gov, and the Oklahoma State University Environmental Health and Safety
information website at http://www.pp.okstate.edu/ehs/fujita.htm)

F-0 Gale tornado (40-72 mph)
Light damage. Some damage to chimneys; branches broken off trees; shallow-rooted
trees pushed over; sign boards damaged.

F-1 Moderate tornado (73-112 mph)
Moderate damage. Peels surface off roofs; mobile homes pushed off foundations or
overturned; moving autos blown off roads.

F-2 Significant tornado (113-157 mph)
Considerable damage. Roofs torn off frame houses; mobile homes demolished; boxcars
overturned; large trees snapped or uprooted; light-object missiles generated; cars
lifted off ground.

F-3 Severe tornado (158-206 mph)
Severe damage. Roofs and some walls torn off well-constructed houses; trains
overturned; most trees in forest uprooted; heavy cars lifted off the ground and thrown.

F-4 Devastating tornado (207-260 mph)
Devastating damage. Well-constructed houses leveled; structures with weak foundations
blown away some distance; cars thrown and large missiles generated.

F-5 Incredible tornado (261-318 mph)
Incredible damage. Strong frame houses leveled off foundations and swept away;
automobile-sized missiles fly through the air in excess of 100 meters (109 yards); trees
debarked; incredible phenomena will occur.

F-6 to F-12 (319 mph to Mach 1)
Inconceivable damage. The maximum wind speeds of tornadoes are not expected to
reach F6 levels.

_A494_

B. Using a truck-mounted Doppler radar system, scientists measured the rotational wind speed of a tornado that struck
Oklahoma City in May, 1999. This is 1 mph less than that
required for a tornado to be classified with a Fujita rating of
F-6 (maximum severity). (*See* table above.)

_A495_

Thunder travels at the speed of sound, which is much slower
than the speed of light. The initial sound of thunder is generated
by that part of the lightning bolt closest to the observer, while
the last sound of a thunderclap is made by the most distant
part of the lightning bolt. Echoing plays a minor role, depending
on the nature of the terrain.

____ A496

True. Fire-induced atmospheric moisture and convection can produce large thunderstorms (colloquially known as pyrocumulus) above the fire. Unfortunately, they usually drift downwind and do not contribute to extinguishing the fire.

____ A497

Downbursts affect areas of up to 15 miles, while microbursts are smaller and affect surface areas of one mile or less in diameter.

____ A498

No airplane is certified to fly in moderate freezing rain.

____ A499

True. Animals have been picked up by tornadoes, lifted into the bellies of thunderstorms, and then deposited along with rain some distance away. More common, however, is when it rains frogs and turtles.

____ A500

128.6° Fahrenheit below zero (-89.2° Celsius) at Vostok Station, Antarctica on July 21, 1983.

____ A501

D. On that same date, tropical cyclone _Denise_ also produced 45 inches of measured rainfall during a 12-hour period.

____ A502

False. There are also needle- or spike-shaped snowflakes; all the rest are six-sided. Although hypothesized, it has not been proven that no two snowflakes are identical in size and shape. Snowflake design, by the way, is determined primarily by the temperature and altitude at which it is formed.

A503

Lightning is headed his way. The pilot has a second or two to immediately drop to the ground in a ball and make as small a target of himself as possible so that the bolt will hopefully strike something taller (such as the aircraft itself). It would be safer to jump into the aircraft, but there probably would not be enough time to do this.

A504

The doldrums are equatorial areas between the trades of the Northern and Southern Hemispheres where the wind is either light and variable or calm, conditions that caused problems for commercial sailing vessels.

A505

Wear a raincoat and carry an umbrella. A hydrometeor is any form of liquid or ice in the atmosphere (including clouds).

A506

The sun heats the surface of the Earth, which in turn heats the atmosphere from below. The higher we climb above this heat source (the Earth) and distance ourselves from it, the more the temperature decreases (approximately 3.5° Fahrenheit, or 2° Celsius, per 1,000 feet in the troposphere).

A507

Glory. Much rarer is a *double glory*, when a second rainbow-like ring concentrically surrounds the first. In German, a glory is called a *heilegenschein*, which literally means halo shine.

A508

True. Also, the worst turbulence is found between the middle and downwind side of the gorge.

A509

136°F (58°C) at Azizia, Libya, in 1922. Although the elevation there is only 380 feet MSL, the density altitude there on that day was 5,100 feet. The hottest temperature ever recorded in the United States was 134°F at Death Valley, California, in 1913.

A510

B. These northeasterly and southeasterly winds meet in a global band called the intertropical convergence zone (ITCZ), which results in an equatorial band of thunderstorms.

A511

True. Low-level winds in the Northern Hemisphere shift right (clockwise) with a gain in altitude. In this case, the wind at the lower altitude is from the north and shifts to a northeasterly wind as altitude increases.

A512

A. Also, low-pressure (or cyclonic) areas typically travel across the United States at about 500 or more statute miles per day.

A513

B. The bases of middle clouds (altocumulus and altostratus) are found between 6,500 feet and 23,000 feet. Towering cumulus and cumulonimbus clouds typically have bases between 1,000 and 10,000 feet AGL.

$$\underline{\hspace{7cm}} \text{A}514$$

It takes a month or two for the Earth to absorb and begin to radiate heat received by the increasing sunlight. Similarly, the warmest month is 1–2 months after the longest day of the year (about June 21st), and the hottest time of day is an hour or two after the sun passes its zenith.

$$\underline{\hspace{7cm}} \text{A}515$$

A. A cyclone is simply a low-pressure area. (An anti-cyclone is a high.) The other phenomena are characterized by violent winds, whereas a cyclonic wind could be as gentle as a zephyr. Take partial credit if you selected C because a tornado is a local phenomenon while the others affect large areas.

CHAPTER 5
AIRCRAFT

Mix 'n' Match the aircraft on the left with the most appropriate clue on the right:

1.______	Bellanca Citabria	A. no rudder pedals
2.______	Piper Tri-Pacer	B. wings move
3.______	Ercoupe	C. geared engine
4.______	Aerostar 601	D. fixed-gear twin
5.______	Helio Courier H-395	E. vertical fin moves
6.______	Cessna 175 Skylark	F. 7GCAA
7.______	Cessna 195	G. turbocharged
8.______	Beechcraft A35 Bonanza	H. radial engine
9.______	Champion Lancer	I. flying milk stool
10.______	Piper Pawnee	J. 260 horsepower
11.______	Cessna Skyhook	K. 185 horsepower
12.______	Mooney 201	L. bug killer

Q 517

An ornithoper is

    A. a helicopter with more than one set of rotors.
    B. a helicopter with one set of rotors.
    C. a gyroplane (or gyrocopter).
    D. an airplane with flapping wings.

Q 518

The Consolidated B-24 Liberator, Cessna 150/152, Cessna 172, and Messerschmitt Bf.109 were produced in significant numbers. Which of the following shows the proper sequence (most produced aircraft first and least produced aircraft last)?

    A. B-24, 150/152, 172, Bf.109
    B. Bf.109, 172, 150/152, B-24
    C. 172, Bf.109, 150/152, B-24
    D. 150/152, 172, B-24, Bf.109

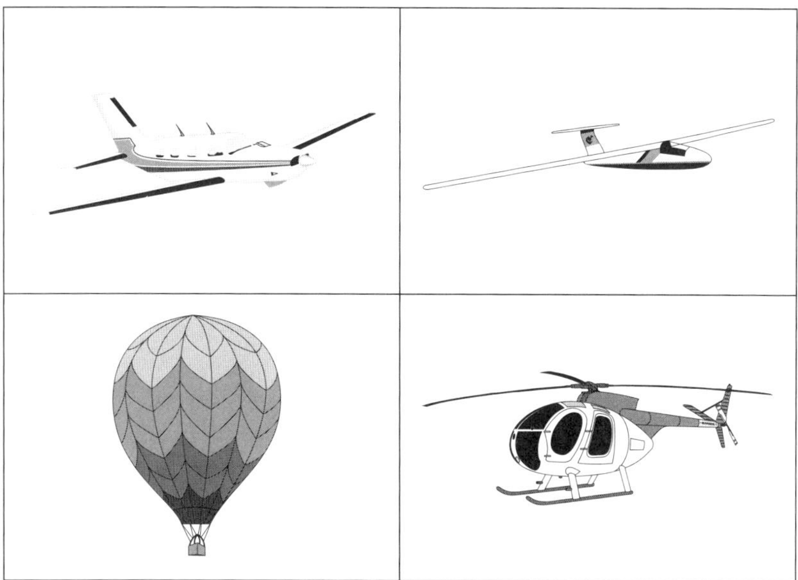

(Q519)

Which of the following does not belong?

    A. airplane
    B. glider
    C. balloon
    D. rotorcraft

True or False? The Douglas DC-3 was also known as the Soviet Lisunov Li-2.

True or False? The Convair 880 jetliner was so named because of its maximum cruising speed.

Questions

Mix 'n' Match the following Cessna models with their appropriate names.

1. Cessna A152 __________	A. Aerobat
2. Cessna 172 __________	B. Agwagon 230
3. Cessna R172E __________	C. Bird Dog (L-19)
4. Cessna FR172E __________	D. Caravan I
5. Cessna FR172K __________	E. Cardinal
6. Cessna 175A __________	F. Centurion
7. Cessna 177 __________	G. Cutlass RG
8. Cessna 182A __________	H. Hawk XP (195 hp)
9. Cessna 185 __________	I. Reims Rocket (made in France)
10. Cessna 188 __________	J. Skylane
11. Cessna 206 __________	K. Skylark (geared engine)
12. Cessna 208 __________	L. Skywagon
13. Cessna 210D __________	M. Super Skywagon
14. Cessna 305A __________	N. T-41B

The Cessna L-19 Bird Dog saw considerable service in the Korean War.   (Q522)

True or False? If the Cessna 310 has two engines, then the Cessna 620 had four.

True or False? Cessna built a Seneca before Piper did, and Piper built a Skyhawk before Cessna did.

Can you name ten 3-engine airplanes?

What unlikely individual was the only person ever to privately own and operate a Douglas DC-5?

What was America's first production turboprop-powered airplane?

Pilots, controllers, and mechanics refer to various aircraft using descriptive nicknames, not all of which are complimentary. Mix 'n' Match the following nicknames with the appropriate aircraft.

1. Sky Box	A. Boeing 727
2. Texas Tube	B. Boeing 737
3. Whale	C. Boeing 747
4. Guppy	D. Britten-Norman Tri-Islander
5. Slobovian DC-10	E. Douglas DC-9
6. Rice Rocket	F. Fairchild Republic A-10 Thunderbolt
7. Jurassic Jet	G. General Dynamics F-16 Fighting Falcon
8. Wart Hog	H. Mitsubishi MU-2
9. Lawn Dart	I. Shorts 330/360
10. Miss Piggy	J. Swearingen Merlin

Most pilots have heard of the Boeing 707, 727, 737, 747, 757, 767, and 777. Prior to Boeing's acquisition of McDonnell-Douglas and its renaming the MD-95, what happened to the Boeing 717?

True or False? Modern two-, three- and four-engine jetliners are equipped with three, four, and five jet engines, respectively.

Can you name the first production, all-metal airplane manufactured by Beech? Cessna? Mooney? Piper?

What is the difference between a sailplane and a glider?

Which of the following does not belong?
- A. cantilever
- B. decalage
- C. interplane strut
- D. stagger

What two airplanes are tied for having the longest, uninterrupted production runs?

In the television series, *Sky King*, the hero, Schuyler "Sky" King, flew a
- A. Beechcraft Model 18 "Twin Beech."
- B. Cessna T-50 "Bamboo Bomber."
- C. Cessna 310.
- D. Piper Apache.

True or False? America's first production business jet was the Learjet.

Mix 'n' Match the following Piper aircraft models with their appropriate names.

1. Apache/Aztec _______	A. PA-18
2. Archer II _______	B. PA-20
3. Arrow III _______	C. PA-22
4. Cherokee, Cruiser, Flite Liner _______	D. PA-23
5. Cheyenne _______	E. PA-24
6. Comanche _______	F. PA-28-140
7. Dakota _______	G. PA-28-151
8. Malibu, Malibu Mirage _______	H. PA-28-181
9. Navajo, Chieftain _______	I. PA-28-236
10. Pacer _______	J. PA-28R-201
11. Seminole _______	K. PA-30
12. Seneca _______	L. PA-31
13. Super Cub _______	M. PA-34
14. Tomahawk _______	N. PA-38
15. Tri-Pacer, Colt _______	O. PA-42
16. Twin Comanche _______	P. PA-44
17. Warrior _______	Q. PA-46

True or False? The 1903 Wright *Flyer* was the first airplane equipped with tricycle landing gear.

One of the most famous airplanes in the world is the Piper J-3 Cub. What does the "J" stand for?

The letters *DC* in DC-3, DC-10, and so forth, stand for

   A. Douglas Civilian.
   B. Douglas Commercial.
   C. Douglas Company.
   D. Douglas Corporation.

Mix 'n' Match the following Beech aircraft models with their appropriate names:

1.______	Model D17S	A.	Baron
2.______	Model D18	B.	Bonanza (aerobatic)
3.______	Model 23	C.	Bonanza (stretched)
4.______	Model C24R	D.	Bonanza (V-tail)
5.______	Model 33	E.	Debonair
6.______	Model F33C	F.	Duchess
7.______	Model V35B	G.	Duke
8.______	Model A36	H.	King Air
9.______	Model D45	I.	Mentor
10.______	Model C50	J.	Musketeer
11.______	Model B55	K.	Sierra
12.______	Model B60	L.	Skipper
13.______	Model 76	M.	Staggerwing
14.______	Model 77	N.	Twin Beech
15.______	Model C90	O.	Twin Bonanza

Arrange the following aircraft in the sequence in which they were originally introduced:

Beech Bonanza Model 35

Cessna 150

Cessna 172

Piper Cherokee 140

In terms of the number of aircraft built, what is the world's most popular jetliner?

_____________________________ **Q 544**

True or False? The Lockheed Constellation has three short vertical stabilizers instead of one tall one. The primary reason for this
was so that a rudder would be directly behind each of the two
inboard engines, and that this would increase rudder effectiveness
in the event of the failure of an outboard engine.

_____________________________ **Q 545**

What was the world's first certified helicopter?

_____________________________ **Q 546**

What is the primary difference between a gyroplane and a helicopter?

_____________________________ **Q 547**

What was the first U.S. airplane to be equipped with counter-
rotating propellers?

_____________________________ **Q 548**

The first production, pressurized, piston-powered, general aviation twin was the

   A. Aero Commander 720 Alti-Cruiser.
   B. Beech Queen Air 88.
   C. Cessna 421 Golden Eagle.
   D. Piper PA-31P Navajo.

_____________________________ **Q 549**

In the classic Humphrey Bogart film, *Casablanca*, the airplane
boarded by Ilsa (Ingrid Bergman) and Victor (Paul Henreid) at
the end of the movie was a

   A. Beech Model 18 Twin Beech.
   B. Lockheed 12A Elektra.
   C. Lockheed 18 Lodestar.
   D. It was not a real airplane.

A four-place airplane with a normally aspirated engine that can cruise at more than 1 mph per horsepower is particularly efficient. What was the first production airplane that could do this?

True or False? The first certified, civilian, production airplane with tricycle landing gear was the Ercoupe.

The first commercial jetliner to be intentionally flown in excess of Mach 1.0 (the speed of sound) was a

    A. Boeing 707.
    B. Convair 990.
    C. Douglas DC-8.
    D. Lockheed 1011.

The only airplane in the world equipped to fly into the core of a mature thunderstorm is

    A. a Beech King Air.
    B. an L188C Electra sailplane.
    C. a Lockheed EC-130Q Hercules.
    D. a North American T-28 Trojan.

Why was America's first jetliner referred to as a 707?

What was the first single-engine, piston-powered airplane capable of cruising at more than 2 mph per horsepower?

Art and Al originally manufactured their airplanes in Wichita, Kansas, and later developed a popular airplane that in 1977 was selected as the "Airplane of the Year" by *Plane and Pilot* magazine. What was their last name?

All four-place Mooney aircraft are designated as model M-20s (such as the M-20J, M-20M, and so forth). Mooney also manufactured three other aircraft, the M-10, the M-18, and the M-22. Can you identify them?

What was the name of Cessna's 8-place, piston-powered single?

Beechcraft's first V-tailed airplane was the
   A. Model 35 Bonanza.
   B. Model 35A Bonanza.
   C. Model V35 Bonanza.
   D. None of the above.

The most powerful piston engine ever built for an aircraft was manufactured by
   A. Lycoming.
   B. Packard.
   C. Pratt-Whitney.
   D. Rolls-Royce.

True or False? The Boeing 747 was the first Boeing airliner to incorporate a spiraling staircase in the cabin.

Match each of the following famous aircraft with the most appropriate clue:

1. *Akron*	A. Airborne aircraft carrier
2. *Albatross*	B. Brrr!
3. *Columbine*	C. Democrat
4. *Enola Gay*	D. His employer's daughter
5. *Glamorous Glennis*	E. His mother
6. *Hindenburg*	F. His wife
7. *Independence*	G. Fictional
8. *Josephine Ford*	H. Nineteen crashes
9. *Memphis Belle*	I. Republican
10. *Spirit of St. Louis*	J. Ryan
11. *Vin Fiz*	K. 25
12. *Winnie Mae*	L. Smoking not allowed

What was a "DC-2 and a half?"

True or False? When viewed from above, the rotors of American- and British-made helicopters rotate in a counterclockwise direction.

Name the specific aircraft type represented by each of the following (unofficial) nicknames:

1. Beechcraft Cherokee _______________________
2. Blackbird _______________________
3. Blue Canoe_______________________
4. Dragon/Shady Lady _______________________
5. Flying Bathtub_______________________
6. Flying Brickyard _______________________
7. Fork-Tailed Devil_______________________
8. Huey_______________________
9. Jug _______________________
10. Mixmaster _______________________
11. Plywood Bullet_______________________

12. Rhapsody in Glue _____________________
13. Spruce Goose_____________________
14. Staggerwing _____________________
15. T-Bone _____________________
16. Tin Goose_____________________ _____
17. Vibrator _____________________
18. Zero _____________________

Q 566

What popular, post-World War II airplane had specific models known as the Standard, the Trainer, the Inter-City Commuter, and the Patroller?

Q 567

Gyroplanes are similar to helicopters in that both incorporate rotors that develop lift. In no-wind conditions, most certified gyroplanes _______ take off vertically and _______ land vertically (no ground roll).

    A. can, can
    B. can, cannot
    C. cannot, can
    D. cannot, cannot

Q 568

In what light, general aviation airplane does the pilot use the control wheel to steer the nosewheel on the ground in the same manner that he uses the steering wheel in an automobile?

Q 569

True or False? The Vought F4U Corsair of World War II fame has a gull wing.

Q 570

True or False? Howard Hughes' *Spruce Goose* is made of spruce.

Q 571

True or False? An antique airplane is defined as any airplane built before World War II.

Q 572

True or False? The first airplane used in airline service that had both supercharging and retractable landing gear was the Douglas DC-3.

Q 573

How did the Piper Cub come to be called a Cub?

Q 574

What was the world's first pressurized airliner?

Q 575

Name a type of well-known aircraft in which the pilot can extend (lower) the landing gear from the wheel wells but cannot retract it.

Q 576

Arrange the following in order of their length, longest first, shortest last.

   A. *Graf Zeppelin* (German airship)
   B. *HMS Titanic* (ocean liner)
   C. *Hindenburg* (German airship)
   D. *USS Akron* or *Macon* (Goodyear-built airships)

Q 577

A non-rigid airship maintains its shape because of the gas pressure within its envelope. It has no internal structure. How did such an aircraft come to be called a blimp?

Only 10 civilian aircraft were certified during World War II. After the war, however, there was a flurry of certification activity. Which of the following aircraft was the first to receive its approved type certificate after the war?

A. Aeronca 7AC Champion
B. Beech 35 Bonanza
C. Ercoupe 415
D. Piper J-3 Cub

The greatest number of engines ever used to power a production airplane was

A. 8.
B. 10.
C. 12.
D. 14.

The most famous of all airships was the *Hindenburg*. Who or what was Hindenburg?

True or False? A seaplane has been safely and intentionally flown in excess of Mach 1.0 (the speed of sound).

True or False? A tandem airplane has two seats, one behind the other.

Airframe manufacturers of yore created some cryptic model designations for their aircraft. Among such aircraft were the Meyers OTW and the Howard DGA. What do these letters signify?

Helium provides buoyancy and maintains the shape of a blimp (as it does a child's balloon). As a blimp climbs, helium — like any gas — expands. What prevents the expanding helium from excessively pressurizing and damaging a blimp? (Note: helium is not dumped overboard because it is expensive and is needed during the descent when gas contracts.)

__________________________________ Q 585

The pilot of a Lockheed SR-71 Blackbird is performing a pre-flight inspection and notes that fuel is dripping incessantly from all six fuel tanks. Why is this considered normal?

__________________________________ Q 586

True or False? When a Mooney pilot applies nose-up trim, the vertical stabilizer simultaneously moves (sweeps) forward.

__________________________________ Q 587

With respect to seaplanes, what is the difference between *pontoons* and *floats*?

The vast majority of floatplanes have high wings to facilitate docking. (Q587)

A canard aircraft is one on which the horizontal stabilizer/elevator is mounted ahead of the wing and near the nose of the aircraft. Why is such an aircraft referred to as a canard?

Which of the following single-engine airplanes have been converted into twins?

A. Ercoupe
B. North American Navion
C. North American P-51 Mustang
D. Piper Super Cub
E. Piper Tri-Pacer

Although helicopters are extremely maneuverable, their performance is limited. Estimate within 10 knots and 5,000 feet the official world speed and altitude records for a helicopter.

True or False? The Lockheed P-38 Lightning was one of the first twin-engine airplanes to have counter-rotating propellers. When viewed from behind, the left propeller turned clockwise, and the right engine turned counterclockwise.

True or False? When a pilot makes a crosswind landing in a Boeing B-52 Stratofortress, his goal is to touch down in a wings-level crab.

What is or was the world's fastest, highest-flying airplane?

____ **Q 594**

____________ manufactured Skymasters at the beginning of
World War II, but _____________ did not produce them until
1961.

____ **Q 595**

True or False? During cruise flight, the canard surfaces ("mous-
tache") of a Beech Starship are swept back 30 degrees, but when
the aircraft is slowed to its landing configuration, they are reposi-
tioned to 4 degrees of forward sweep.

____ **Q 596**

Which of the following is not true?
   A. The 172 was Cessna's first production airplane equipped
      with tricycle landing gear.
   B. The Hawker Siddeley Trident was the first production airliner
      with an autoland system.
   C. On early model Boeing B-52 Stratofortresses, the tail gunner
      could escape the aircraft by jettisoning the entire tail turret.
   D. The pilot of the mammoth, single-engine biplane, the
      Antonov AN-2 Colt (*Anushka* or *Little Anna*) can use an on-
      board compressor to vary the air pressure in the oleo struts
      to suit the type of surface on which he intends to land.

____ **Q 597**

True or False? All blimps and Zeppelins are airships and dirigibles.

____ **Q 598**

True or False? The Concorde supersonic transport has wing flaps.

____ **Q 599**

True or False? The huge Consolidated B-36 Peacemaker was said
(with respect to its engines) to have "six turning and four burning."
In other words, it had six turboprop and four turbojet engines.

Which of the following does not belong?

A. Kachina
B. Morrisey
C. Shinn
D. Varga

True or False? Blimp manufacturer Goodyear also produced inflatable airplanes.

1F, 2I, 3A, 4G, 5J, 6C, 7H, 8K, 9D, 10L, 11B, 12E.

Cessna's first and only helicopter was the Skyhook.   (A516)

D. Early pioneers attempted to build airplanes with flapping
   wings, but none were successful.

C. As of August 1, 2009, Cessna had manufactured more than 46,000 172s (and climbing). The Messerschmitt Bf.109 is next at "close to" 35,000 followed by 28,942 Cessna 150/152s, and 18,188 B-24 Liberators (more than any other U.S. military aircraft).

C. Airplane, glider, and rotorcraft are categories of aircraft. The fourth category of aircraft is lighter-than-air, not balloon. (Balloons and airships are members of the lighter-than-air category.)

True. The Soviet Union had a licensing agreement during World War II to build the DC-3.

True. The Convair 880 could fly at 880 feet per second (521 knots).

1. A		8. J	
2. N		9. L	
3. I		10. B	
4. H		11. M	
5. G		12. D	
6. K		13. F	
7. E		14. C	

True. The Cessna 620 was a 10-seat, pressurized, executive aircraft with four 320-hp engines. Only one was built, however.

Cessna 620.   (A523)

True with respect to the Seneca and false with respect to the Skyhawk. The Cessna Seneca was the military version of Cessna's CH-1B/YH-41 helicopter.

There are many more, but here are 10 that come to mind:

Boeing 727	Ford Tri-Motor
Britten-Norman Tri-Islander	Junkers Ju.52
Dassault Falcon 50	Hawker Sideley Trident
Douglas DC-10	Lockheed L-1011 Tri-Star
Fokker Trimotor	McDonnell Douglas MD-11

Another three-engine airplane is the Bushmaster 2000, a derivative of the Tri-Motor Ford.   (A525)

William Boeing, *the* William Boeing.

The Lockheed C-130 Hercules.

Answers

1. I   6. H
2. J   7. E
3. C   8. F
4. B   9. G
5. D   10. A

Prior to the acquisition of McDonnell Douglas by Boeing, the model 717 was Boeing's in-house (informal) designation for the U.S. Air Force KC-135 Stratotanker (used for air-to-air refueling) and C-135 military transport. Subsequent to the acquisition, Boeing renamed the MD-95 as the Boeing 717.

True. The additional jet engine is the auxiliary power unit (APU) that provides electrical and pneumatic power primarily when the aircraft is on the ground.

Beechcraft Model 35 Bonanza, Cessna 195 (which preceded the Cessna 190), Mooney M.20B Mark 21, and Piper PA-23 Apache.

According to the Soaring Society of America, there is no difference. However, common usage implies that a sailplane has better glide performance than a glider.

A. The other items refer exclusively to the features of a biplane.

They are the Beechcraft Bonanza and Yakovlev Yak-18, both of which have been in continuous production since 1947 (as of the date of this publication).

B. and C. The T-50 was used in early episodes and the 310 in later ones.

Sky King's first airplane was a Cessna UC-78 Bobcat (Bamboo Bomber). (A535)

False. A production model of the Lockheed 1329 Jetstar first flew during the summer of 1960, which preceded the Learjet by more than three years.

____ A537

1. D	10. B
2. H	11. P
3. J	12. M
4. F	13. A
5. O	14. N
6. E	15. C
7. I	16. K
8. Q	17. G
9. L	

____ A538

False. The Wright *Flyer* was equipped with a pair of long skids.

____ A539

The letter *J* represented Walter C. Jamouneau, William Piper's chief engineer during the evolution of the J-2 and J-3 Cubs.

____ A540

B. Technically, the DC-3 is the Douglas Commercial [Model Number] 3, and so forth.

____ A541

1. M	9. I
2. N	10. O
3. J	11. A
4. K	12. G
5. E	13. F
6. B	14. L
7. D	15. H
8. C	

Beech Bonanza (1947)
Cessna 172 (1956)
Cessna 150 (1959)
Piper Cherokee (1961)

The Boeing 737. Until 1987, the Boeing 727 held this distinctive title.

False. Although this sounds logical and might be an indirect result of the design, it was not the reason for the Connie's distinctive tail. The real reason for the triple-tail is that Howard Hughes advised Lockheed that he would not purchase the Constellation for Trans World Airlines unless the airplane would fit completely into the airline's existing hangars, and these were not tall enough to accommodate a single vertical fin.

Lockheed Constellation abeam the southern tip of Manhattan Island.   (A544)

_______________________________________ **A545**

The Bell 47, which was certificated in 1947.

_______________________________________ **A546**

The rotor of a helicopter is powered; the rotor of a gyroplane (autogyro) is not powered and turns by autorotation.

_______________________________________ **A547**

The Wright *Flyer* (1903). The first U.S. production airplane appears to have been the Lockheed P-38 Lightning.

_______________________________________ **A548**

A. The Alti-Cruiser was first produced in 1958. The others were introduced in 1966, 1967, and 1970, respectively.

_______________________________________ **A549**

B. It was the same basic type used by Amelia Earhart during her ill-fated attempt to fly around the world.

_______________________________________ **A550**

The Mooney Mark 20, which first flew on August 10, 1953 and cruises at more than 150 mph with a 150-hp engine.

_______________________________________ **A551**

False. The first was the Curtiss-Wright amphibious Commuter (1935), which was followed by a few lesser-known aircraft. The Ercoupe (1940) was the first general aviation airplane with tricycle gear to achieve popularity.

_______________________________________ **A552**

C. A Douglas DC-8-43 was flown at Mach 1.012 during a test flight over Edwards Air Force Base in 1961.

$$A553$$

D. The other aircraft are used to fly into *developing* thunder-storms. The special T-28 has armor-plated leading edges, steel bracing over the canopy, and a ¾-inch-thick windshield.

$$A554$$

The model number, 707, was selected simply because it sounded good.

$$A555$$

The Wright *Flyer*, which achieved 30 mph with a 12-hp engine.

$$A556$$

Mooney. The airplane was the Mooney 201.

$$A557$$

The Mooney Cadet (nee Ercoupe), the Mooney Mite (single place), and the Mooney Mustang (pressurized single), respectively.

$$A558$$

The Stationair 8, which was essentially a Cessna 207 Skywagon with eight seats.

$$A559$$

D. The first was a V-tailed version of the Beechcraft AT-10 Wichita, a twin-engine, advanced trainer built for the military during World War II.

$$A560$$

A. Lycoming's XR-7755 was a 36-cylinder, radial engine that developed 5,000 horsepower and weighed 6,050 pounds. It was never used on a production airplane.

Answers

A561

False. The Boeing Clipper (Model 314), a flying boat, had a spiraling staircase, as did the Boeing Stratocruiser (Model 377), which was developed from the military C-97.

A562

1. A. A mammoth, Goodyear dirigible that could launch and retrieve airplanes.
2. G. Jules Verne's huge helicopter as described in Clipper of the Clouds.
3. I. President Dwight Eisenhower's Lockheed Constellation.
4. E. Colonel Paul Tibbets, Jr. flew this B-29 to Hiroshima.
5. F. Charles "Chuck" Yeager's Bell X-1.
6. L. Last dirigible to use hydrogen.
7. C. President Harry Truman's Douglas DC-6.
8. B. Admiral Richard Byrd's plane to the North Pole.
9. K. Boeing B-17 that escaped unscathed after 25 bombing missions over enemy territory.
10. J. Built by Ryan Airlines in San Diego.
11. H. First flight across the United States (Calbraith Rodgers).
12. D. Wiley Post flew twice around the world in this Lockheed Vega.

A563

A China National Airways Corporation DC-3 was strafed on the ground near Kiuchuan, China by Japanese fighters in 1941. The right wing was destroyed and replaced by a DC-2 wing that was 10 feet shorter. Somewhat askew the airplane flew.

A564

True. The rotors of Russian- and French-made helicopters, however, rotate in a clockwise direction.

1. Beech Model 23 Musketeer
2. Lockheed SR-71
3. USAF L-27 (militarized Cessna 310)
4. Lockheed U-2
5. Aeronca C-2
6. Space Shuttle (referring to the heat-resistant tiles)
7. Lockheed P-38 Lightning
8. Bell UH-1 helicopter (from the early HU-1 designation)
9. Republic P-47 Thunderbolt (originally named Thunderjug)
10. Cessna 337 Skymaster (a.k.a. "Push Me, Pull You")
11. Lockheed Vega
12. Cessna T-50 (USAAF's wooden UC-78 Bobcat)
13. Hughes-Kaiser HK-1 Hercules flying boat
14. Beech Model 17 biplane (a.k.a. the Flying Backstagger)
15. Beech Model 50 Twin Bonanza
16. Ford 4-AT and 5-AT Tri-Motor (from Ford's "Tin Lizzy" automobile)
17. Vultee BT-13 Valiant (from its substantial stall-warning buffet)
18. Mitsubishi A6M fighter (Allied name "Zeke")

The Cessna 150. The Cessna A150 is known as the Aerobat.

B. Engine power is used to accelerate the rotors of most certified gyroplanes while the aircraft are at rest. The pilot then disengages a clutch-type mechanism and uses rotor momentum to make a jump takeoff. Although they can land in a very short distance, none can land without some ground roll.

____ A568

The Ercoupe. Because it has no rudder pedals — the ailerons and rudders are interconnected — the control wheel must be used for ground steering. (Crosswind landings are made while crabbing into the wind.)

____ A569

False. The Corsair has an *inverted* gull wing.

____ A570

False. The *Spruce Goose* is predominantly birch.

____ A571

False. An antique airplane is one built before the end of 1945.

____ A572

False. The first such airplane was the Lockheed Orion.

____ A573

The airplane was originally powered by a 20-hp, two-cylinder Brownback "Tiger Kitten" engine. Because a tiger kitten is a cub, the company's accountant, Gilbert Hadrel, was inspired to call the little airplane a Cub.

____ A574

The Boeing 307 Stratoliner, a four-engine taildragger that first flew on the last day of 1938.

____ A575

Any of NASA's five space-shuttle orbiters: *Atlantis*, *Challenger*, *Columbia*, *Discovery*, and *Endeavor*. The same is true of the research vehicle, *Enterprise*, but it was not capable of orbital flight.

Titanic (883 feet)
Hindenburg (803 feet)
Akron/Macon (785 feet)
Graf Zeppelin (775 feet)

During World War I, Lt. A. D. Cunningham of the Royal Navy Air Service commanded the air station at Capel, England. While walking along the side of His Majesty's (non-rigid) Airship SS-12, he playfully flipped his thumb at the taut fabric "gasbag" and verbalized the odd sound that this made, "blimp." The Lighter-Than-Air Society dismisses as myth the notion that *blimp* was a contraction of "Type B, Limp."

A. The "Champ" (or "Airknocker") was certified on October 18, 1945, and by August, 1946, 43 of them were rolling out of the factory every day. The Cub and the Ercoupe were certified before the war, and the Bonanza was certified on March 25, 1947. (The Department of Commerce issued Approved Type Certificate No. 1 to the Buhl-Verville J4 Airster on March 29, 1927.)

An Aeronca 7DC Champion is the same as a Model 7AC Champion, except that it has a larger vertical fin and an extra 10 horsepower.   (A578)

C. The 1929 German Dornier Do X flying boat was powered by 12 Curtiss Conqueror engines mounted in six pairs along the top of the wing. Each pair consisted of a tractor and a pusher mounted fore and aft in a single nacelle.

Paul von Hindenburg was the second president of the Weimar Republic and appointed Adolf Hitler chancellor of Germany in 1933. The mammoth airship was to have been named the *Hitler*, but *der Fuehrer* reportedly would not permit his name to be used on anything that could fail or be destroyed.

True. The Convair XF2Y-1 Sea Dart was similar to Convair's F-102 Delta Dagger, but instead of wheels, it was equipped with retractable water skis. After entering a shallow dive on August 3, 1954, the Sea Dart became the world's first and possibly only supersonic seaplane.

False. That is a tandem-seated airplane or an airplane with tandem seats. A tandem airplane has two (or more) wings with one behind the other in approximately the same plane.

OTW stands for *out to win*, and DGA stands for *darn good airplane*.

There are two large air-filled bags, called *ballonets*, inside the blimp. During climb, air is ported overboard from the ballonets, which then contract to make room for the expanding helium. When the helium contracts during descent, ram-air pressure refills the ballonets to maintain constant helium pressure.

Components of the SR-71 (including its fuel tanks) fit loosely when the aircraft is on the ground. At cruise speeds of 2,000 knots, air friction increases skin temperature to as high as 600 degrees Fahrenheit. Such heating causes components to expand, thereby sealing the fuel tanks and preventing in-fight leakage.

True. The trimmable horizontal stabilizer and the vertical stabilizer move as a single unit in response to pitch-trim input. Similarly, the vertical fin moves (sweeps) aft when nose-down trim is applied. This movement of the vertical stabilizer presumably improves rudder effectiveness at low speed and reduces drag at high speed.

There is no difference. *Pontoon* is an obsolete word used to describe a float.

Canard is French for *duck*. The aircraft is so named because the aft location of the wing makes it look like a duck in flight. (A duck's wing also is at the rear of its body.) The 1903 Wright *Flyer* was a canard design.

All of them. They were, respectively, the Erco Twin Ercoupe, Riley Twin Navion, North American F-82 Twin Mustang, Wagner/Piper Twin Cub, and Wagner Twin Tri-Pacer. (Technically, the F-82 was not a direct conversion *per se*; it was a conversion of the P-51 design.)

_______________________________ A590

216.46 knots in a Westland Lynx (British) and 40,820 feet in
an Alouette SA 315-001 Lama (French). Both aircraft are
turbine powered.

_______________________________ A591

False. This would have been the preferred arrangement to mini-
mize V_{MC} (minimum controllable airspeed with an engine out).
The P-38's propellers, however, turned contrarily. The primary
purpose of counter-rotating the propellers in this manner was
to improve the Lightning's spin characteristics.

_______________________________ A592

True. The bicycle-type landing gear is hydraulically aligned with
the runway (by a crewmember) prior to touchdown so that there
is no side loading on the tires when the aircraft touches down
in a crab.

_______________________________ A593

The rocket-powered North American X-15 was launched from a
"mother" aircraft, achieved a speed of 4,534 mph (3,940 knots)
and holds an altitude record of 354,330 feet.

_______________________________ A594

The *Douglas* C-54 Skymaster was the military variant of the four-
engine DC-4, and *Cessna's* versions were the "push-me, pull-you"
models 336 (fixed landing gear) and 337 (retractable gear).

_______________________________ A595

True. This helps to offset the nose-down pitching moment
created by flap deployment.

A596

A. The Cessna 310 was first produced in 1954. The prototype Cessna 172 did not fly until 1955.

A597

True. An airship is any lighter-than-air aircraft that is powered and steerable. A Zeppelin is any of many rigid airships designed by Count Ferdinand von Zeppelin. A blimp is a non-rigid airship. The terms *airship* and *dirigible* (from directable) are synonymous.

A598

Concorde does not have wing flaps *per se*, but its six elevons droop for takeoff and landing to increase wing camber. Effectively, therefore, Concorde has flaps.

A599

False. The mammoth 10-engine, intercontinental bomber of the Cold War era did not have any turboprop engines. It was powered by six 3,800-hp, Pratt & Whitney radial (piston) engines and four 5,200-pound-thrust, General Electric J-47 turbojets.

A600

A. Morrisey, Shinn, and Varga manufactured the model 2150 trainer at different times. Varga, however, called the airplane a Kachina the way Cessna calls the model 172 a Skyhawk.

A601

True. Goodyear Tire and Rubber manufactured several 1- and 2-place Inflatoplanes in the late 1950s. They had inflatable fuselages, wings, and tail surfaces. Unfortunately, the project didn't float.

CHAPTER 6
AIRLINE
&
MILITARY

Stove lids—which are flat, cast-iron discs found on old stoves—
were used by many World War I pilots to

A. place on their seats.
B. drop on the enemy.
C. bribe the enemy after a crash-landing behind enemy lines.
D. adjust an airplane's center of gravity.

Q 603

True or False? A small observation compartment used to be
attached to the belly of some aircraft. It was lowered by cable
during flight so that someone sitting in the compartment could
make weather, terrain, and troop observations in clear air while
the mother ship hid in cloud from the enemy.

Q 604

Spell the name of Australia's largest, most famous airline.

Q 605

Why do pilots refer to anti-aircraft fire as flak?

Q 606

The Boeing B-17 bomber is a four-engine taildragger that does
not have reversible-pitch propellers. How is it possible for a pilot
to back the Flying Fortress into a parking space from within
the cockpit?

Q 607

Which jetliner in each of the following seven categories was the
first to be placed in service?

1. First twin-engine jet transport
2. First three-engine jet transport
3. First four-engine jet transport

...Continued

Continued from page 244

4. First wide-body, twin-engine jet transport
5. First wide-body, three-engine jet transport
6. First wide-body, four-engine jet transport
7. First supersonic jet transport

Q 608

A pilot walks into the cockpit of a typical jetliner and immediately notices the Deperdussin control. What is it?

Q 609

Boeing flew one of its aircraft into the highest elevation airport ever to receive a jetliner. Where is this airport, and how high is it?

Q 610

Why were certain French World War I airplanes called penguins?

Q 611

What is the world's shortest, scheduled airline flight?

Q 612

Specifically, who were Jiro Horikoshi, Edgar Schmued, Willy Messerschmitt, and Reginald Mitchell?

Q 613

Why were some American World War II pilots given extensive training in crash landing techniques and very little instruction in conventional landing techniques?

Q 614

Why is landing a landplane in water called ditching?

Q 615

True or False? During World War II, Germany developed special goggles that allowed *Luftwaffe* fighter pilots to "see" turbulence less than 1.5 miles ahead of the aircraft.

Name two things that are wrong with the following statement:
While in cruise flight at 25,000 feet during World War II, the
P-51 pilot lit a cigarette and settled back to enjoy the flight.

Why did pilots of yore wear silk scarves?

Many World War II pilots gained fame by "flying the hump."
What was the hump?

Famed World War II *Luftwaffe* ace, Adolf Galland, adorned his
fighter with which of the following insignias?
   A. Mickey Mouse
   B. a Star of David
   C. a caricature of Marilyn Monroe
   D. a frankfurter

Why were some Japanese World War II fighters called Zeros?

True or False? During World War II, the swastika adorned the
wings of Luftwaffe fighters.

True or False? During World War II, Germany developed motor-
less fighter gliders.

At any time when Air Force One is not airborne, it can be said that it has made one more takeoff than it has made landings, yet has never crashed. How is this explained?

A bombardier releases bombs from an airplane, but during World War II, what did a bomphleteer do?

In which country did the United States make first use of military airplanes in warfare?

True or False? During World War II, a pilot flying a British dirigible pursued and shot down a Heinkel He.115.

During World War II, which of the following bugs was important to aerial warfare?

- A. hornet
- B. dragonfly
- C. spider
- D. beetle

True or False? Airplanes have been equipped with water skis.

How did the Jenny, the famed trainer of World War I, get its name?

Balloon fenders were used on

- A. airplanes.
- B. balloons.
- C. dirigibles.
- D. helicopters.

Israel's national airline is El Al Israel Airlines. What do the words *El Al* signify?

During World War II, British pilots occasionally dropped calling cards from their aircraft. Why was it dangerous to pick up one?

How did parrots play a role in aerial warfare during World War I?

For what military purposes were kites used during World War II?

Can you estimate within 50 feet the total distance available for takeoff on the deck of the aircraft carrier *Hornet* when the first B-25 took off to spend *Thirty Seconds Over Tokyo* during World War II?

Why were the vertical tail surfaces of Northwest Airlines' aircraft painted solid red (or nearly so)?

____ **Q 637**

The co-inventor of a frequency-switching munitions device that was
to allow pilots to guide torpedoes once they entered the water was

    A. actor Clark Gable.
    B. actress Hedy Lamarr.
    C. General William "Billy" Mitchell.
    D. famed pilot R. A. "Bob" Hoover.

____ **Q 638**

True or False? During World War II, many carrier-based airplanes
were equipped with flotation devices to keep them afloat following a ditching.

____ **Q 639**

Pilots know that the *B* in B-29 indicates that the Superfortress was
designed for bombardment; it is a bomber. What do the letters in
the following World War II aircraft designations indicate?

    1. A-26 Invader   8. OA-10 Catalina
    2. AT-6 Texan   9. P-51 Mustang
    3. BT-13 Valiant   10. PT-23 Cornell
    4. C-47 Skytrain   11. R-6
    5. F-8 Mosquito   12. UC-78 Bobcat
    6. L-4 Grasshopper 13. CG-3A
    7. O-52 Owl

____ **Q 640**

How did Delta Air Lines get its name?

____ **Q 641**

Their official name was the American Volunteer Group (AVG),
but they were better known as _____________________.

____ **Q 642**

What must a pilot do to qualify for membership in the Caterpillar
Club?

The first woman CEO (chief executive officer) of a scheduled airline was

A. Jacqueline Cochran.
B. Leona Helmsley.
C. Maureen O'Hara.
D. Lana Turner.

Many World War II airplanes were all metal except that their control surfaces were covered with fabric. The primary purpose for using fabric on such airplanes was to

A. reduce aircraft gross weight.
B. reduce manufacturing cost.
C. reduce the possibility of flutter.
D. simplify repairs in the field.

A pilot wants to "go for a hop" in his new airplane. How did the word "hop" come to mean "a local flight?"

Which of the following was the first U.S. production airplane equipped with turbocharging or supercharging?

A. Boeing B-17A Flying Fortress
B. Curtis P-40 Warhawk
C. Douglas DC-3/C-47 Skytrain
D. Lockheed P-38 Lightning

(Q647)

____ **Q 647**

True or False? One way to determine that the Boeing 747 used in the movie, *Air Force One*, is not the genuine article is that it lacks the bulge on the nose of the aircraft that conceals the aerial-refueling receptacle.

____ **Q 648**

Mosquito raids during World War II were _______________.

____ **Q 649**

What was the fastest production fighter of World War II?

____ **Q 650**

Speaking of World War II, who was "Chicken" Kamikaze?

____ **Q 651**

What World War II pilot with a now-famous name took off from the aircraft carrier *Lexington* in his Grumman Wildcat and shot down five Japanese bombers in less than five minutes?

____ **Q 652**

True or False? In-flight motion pictures were first used to entertain airline passengers in 1925.

____ **Q 653**

Why did many World War II bombs whistle after being dropped?

____ **Q 654**

True or False? During World War II, the pilot of a Japanese warplane bombed the west coast of Oregon.

____ **Q 655**

True or False? The first use of manned aircraft in the United States for military purposes occurred during the Civil War.

____ **Q 656**

One of the most famous operations in the history of the U.S. Air Force was called *Operation Vittles*. What was its more popular name?

____ **Q 657**

Why were some bombs dropped by parachute during World War II?

____ **Q 658**

What is the aeronautical origination of the expression, "the whole nine yards"?

____ **Q 659**

Why did airmen training during World War II at Midland, Texas, have a higher incidence of black eyes than airmen training anywhere else?

____ **Q 660**

True or False? In military aviation, a sortie usually is a combat mission consisting of two aircraft.

____ **Q 661**

Why was it often desirable during World War II for flight crews to include someone who was colorblind?

—————————————————————————— **Q 662**

When was an aircraft first put to military use, and how was it used?

—————————————————————————— **Q 663**

During World War II, Tokyo Rose broadcast propaganda intended to demoralize Allied airmen and troops. Who was her European counterpart?

—————————————————————————— **Q 664**

In which years did the Boeing B-47 (the XB-47) and the Boeing B-52 (YB-52) make their first flights? (You have already been given sufficient clues.)

—————————————————————————— **Q 665**

Who was America's "Ace of Aces"?

—————————————————————————— **Q 666**

Flat-hatting is a term that originated in military aviation and describes flying at dangerously and unnecessarily low altitudes (usually for the thrill of it). It is synonymous with buzzing. How did this term originate?

—————————————————————————— **Q 667**

What was the name of the group of daring and flamboyant American pilots that was organized by Claire Chennault and operated in China between 1937 and 1941?

   A. Confederate Air Brigade
   B. Flying Tigers
   C. Lincoln Brigade
   D. 14th VBS

—————————————————————————— **Q 668**

True or False? The *Lafayette Escadrille* was a squadron composed of French fighter pilots that were equipped with American aircraft and served under American command during World War II.

____ **Q 669**

Japanese pilots who sacrificed their lives in suicidal missions were called *kamikaze* pilots. In Japanese, *kamikaze* means *divine wind.* Why were these pilots so named?

____ **Q 670**

During World War II, military pilots developed a colorful "slanguage." Define the meanings of the following slang expressions:

1. blanket drill	7. pulpit
2. Chinese landing	8. roll up your flaps
3. flying the iron beam	9. rug dance
4. French landing	10. shot down in flames
5. geese	11. sugar report
6. laying eggs	

____ **Q 671**

True or False? During World War II, some of Germany's jet fighters were equipped with guided missiles.

____ **Q 672**

What was probably the most unconventional means by which a fighter pilot attacked and downed an enemy aircraft?

____ **Q 673**

Grumman Aircraft manufactured eight fighters that were named after cats. How many of them can you name?

____ **Q 674**

What was the first airline to institute a frequent-flier program?

____ **Q 675**

What was the only U.S. fighter aircraft to be in production before, during, and after World War II?

The *Luftwaffe* is the name of Germany's air force. What does the word mean in English?

True or False? During World War II, a Douglas DC-4 was stripped of its engines and converted into a cargo-carrying glider.

During World War II, what were *Weary Willies*?

When a pilot makes an early morning departure, he might refer to leaving on a dawn patrol. What is the origination of the term *dawn patrol*?

The first person to be awarded the Distinguished Flying Cross (DFC) was _______________, and the only civilian ever to receive one was _______________.

How were metal darts used during aerial warfare in World War I?

What was America's first all-metal fighter? (It also was the first monoplane to enter U.S. military service, the last with an open-cockpit, the last with fixed landing gear, and the last with externally braced wings.)

In terms of the number of aircraft shot down, who is the greatest ace of all time? (An ace must have shot down at least five aircraft.)

During World War II, American paratroopers were advised to yell when leaping from their airplanes to relieve pressure on their ears and lessen nervous tension. Why was it popular to yell the word *Geronimo*?

Why is attacking ground personnel or installations by aircraft firing machine guns called *strafing*?

A. Early fighter pilots flew fabric-covered airplanes and sat on stove lids to protect their backsides against enemy ground fire. Stove lids were the first form of armor used in airplanes.

A603

True. Some early dirigibles were equipped with such sub-cloud cars.

A604

There is no "u" in Qantas Airways. Qantas is an acronym for Queensland and Northern Territories Aerial Services.

A605

Flak is an acronym that comes from the German expression *flieger abwehr kanone*, which means "anti-aircraft cannon fire."

A606

The left and right outboard engines are outboard of the left and right main landing gear wheels, respectively (see figure below). The pilot locks the left brake and applies power to the left out-

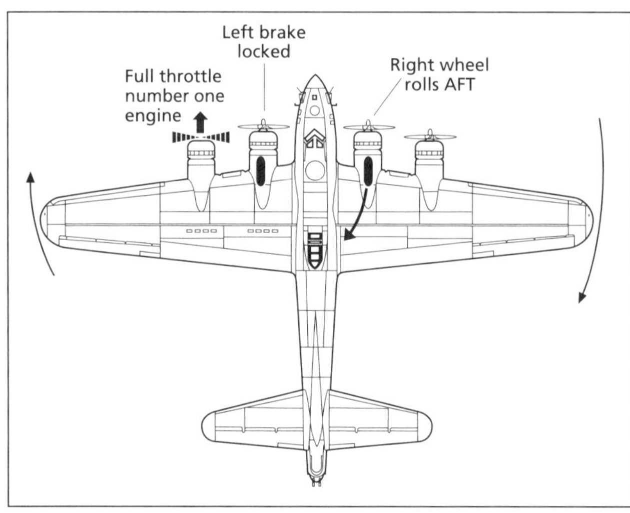

(A606)

Continued from page 257

board engine. This causes the aircraft to pivot on the left tire, which forces the right tire to roll aft. The pilot then applies right brake and adds power to the right outboard engine, which causes the left tire to roll aft. Alternating in this manner, the pilot gradually "walks" the entire airplane backward.

A607

1. Sud-Est Aviation
   S.E. 210 Caravelle
2. Boeing 727
3. De Havilland Comet
4. Airbus Industrie A300
5. Douglas DC-10
6. Boeing 747
7. Tupolev Tu-144

The Boeing 727, world's first three-engine jetliner.   (A607)

A608

A Deperdussin control is the conventional wheel-and-column type of flight control (also called a dep control), which was developed by a Belgian, Armand Deperdussin.

A609

Banda, Tibet, is near Changdu, Tibet and has an elevation of 14,219 feet MSL.

$$A610$$

Penguins were aircraft with extremely short wings and could not fly. They were used only to practice taxiing and takeoff and landing rolls.

$$A611$$

The world's shortest airline flight is between Westray and Papa in the Orkney Islands (Scotland) and takes 2 minutes.

$$A612$$

They were responsible for designing their countries' best World War II fighters: the Mitsubishi A6M2 Zero, the North American F-51 Mustang, the Messerschmitt Bf. 109 (Me. 109), and the Supermarine Spitfire, respectively.

$$A613$$

Glider pilots were trained to make only one landing. This consisted of crash landing their load of commandos behind enemy lines at night and without the help of landing lights.

$$A614$$

Ditching originated as a Royal Air Force term that meant landing in the English Channel, or "the ditch."

$$A615$$

False. It would be convenient, however, if such goggles were available to general aviation pilots.

$$A616$$

World War II fighters were not pressurized, so their pilots had to wear oxygen masks at such high altitudes. Also, there is insufficient oxygen at 25,000 feet for a cigarette to remain lit.

        Answers

$$A617$$

American World War I pilots started the custom. The scarves pre-vented skin irritation that otherwise was caused by rubbernecking (looking behind for enemy aircraft) while wearing heavy, scratchy, uniform shirts (to keep warm at cold altitudes in open-cockpit aircraft).

$$A618$$

The hump was and still is the Himalayan Mountains of southern Asia.

$$A619$$

A. General Galland was a fan of Disney cartoons.

$$A620$$

They were technically known as Zero Zeros, which represented the last two digits of the year in which they were designed: the 2,600th year of the Nipponese Dynasty (1940).

$$A621$$

False. The *swastika* was placed on the vertical stabilizer and/or the rudder.

$$A622$$

True. Although never deployed, the Blohm and Voss 40 was equipped with a 20-mm cannon. After being towed aloft, it was to swoop through allied bomber formations and destroy what it could in the process. If the pilot were to survive such a mission, he would simply glide to a landing.

—————————————————————— A623

When Richard M. Nixon left office, his resignation became effective after departing Andrews Air Force Base on *Air Force One*. Gerald Ford was sworn into office while the aircraft was en route to California. At this moment, Nixon's flight lost its presidential designation and assumed a military flight number.

—————————————————————— A624

He dropped propaganda pamphlets. The word was coined by British pilots making pamphlet raids over Europe.

—————————————————————— A625

Mexico. In 1914, navy airplanes took action against hostile gunfire during the American occupation of the seaport in Vera Cruz.

—————————————————————— A626

False. You've got to be kidding.

—————————————————————— A627

C. Spiders provided web filaments that were used for crosshairs on most sighting apparatus, including bombsights.

—————————————————————— A628

True. In 1950, the navy contracted a Delaware firm to conduct flight tests and sea trials of a North American SNJ (the Navy equivalent of a T-6 Texan) and a Piper PA-11 equipped with water skis (later called hydroskis). The purpose was to make it possible to takeoff from and land on a 120-foot-long raft made of balsa (dubbed the *U.S.S. Balsawood*). Although not practical, the tests were successful.

A629

The aircraft was officially designated as a Curtiss JN-4D, but pilots found it easier and preferred to call it a JN or Jenny.

A630

A. During World War II, balloon fenders were guard rods that extended from the wing tips to the nose of an airplane to protect it from entanglement in the cables of barrage balloons.

A631

El Al is Hebrew for *To the Skies* or *Onward and Upward* (depending on the translator).

A632

English calling cards were small pieces of cardboard treated with yellow phosphorous that burst into flame upon drying.

A633

Because of their acute hearing, parrots positioned on the Eiffel Tower warned of approaching aircraft long before the planes were heard or seen by ground observers.

A634

Russia used kites to drop propaganda leaflets over the eastern front, and the Allies raised them over ships as protection against dive-bombers.

A635

404 feet. The carrier steamed into a strong surface wind to enable the heavily loaded aircraft to become airborne.

$$A636$$

During the airline's infancy, it operated mostly in the northern United States. The tails were painted red to enhance locating a downed aircraft in the snow.

$$A637$$

B. Lamarr (nee Hedwig Eva Maria Kiesler) was much more than a glamorous movie star.

$$A638$$

True. The flotation gear consisted of large rubber bags packed into the wings. The pilot inflated the bags with carbon dioxide from a highly charged flask.

$$A639$$

1. Attack (light bombardment)
2. Advanced trainer
3. Basic trainer
4. Cargo and transport
5. Reconnaissance and photographic
6. Liaison
7. Observation
8. Amphibian
9. Fighter (pursuit)
10. Primary trainer
11. Rotary wing (helicopter)
12. Utility transport
13. Cargo-troop glider

$$A640$$

Delta began life in the Mississippi River Delta in 1925 as the world's first crop-dusting operation.

$$A641$$

The Flying Tigers.

$$\_\_\_\_\_\_\_\_\_\_\_\_\_\_\_\_\_\_\_\_\_\_\_ A642$$

He must have used a parachute to save his life.

$$\_\_\_\_\_\_\_\_\_\_\_\_\_\_\_\_\_\_\_\_\_\_\_ A643$$

C. The famous Hollywood actress was also an aviation pioneer
and headed the Caribbean-based airline, Antilles Air Boats.

$$\_\_\_\_\_\_\_\_\_\_\_\_\_\_\_\_\_\_\_\_\_\_\_ A644$$

C. Fabric reduced the weight of the control surface aft of the
hinge line, which reduced the likelihood of flutter.

$$\_\_\_\_\_\_\_\_\_\_\_\_\_\_\_\_\_\_\_\_\_\_\_ A645$$

The acronym, *hop*, evolved during World War I and meant a
"high operations patrol."

$$\_\_\_\_\_\_\_\_\_\_\_\_\_\_\_\_\_\_\_\_\_\_\_ A646$$

A. The B-17A prototype and the production model first flew on
April 29, 1938 and June 27, 1939, respectively.

$$\_\_\_\_\_\_\_\_\_\_\_\_\_\_\_\_\_\_\_\_\_\_\_ A647$$

True. The "Flying White House" can be refueled in flight.

$$\_\_\_\_\_\_\_\_\_\_\_\_\_\_\_\_\_\_\_\_\_\_\_ A648$$

A mosquito raid was a bombing and/or strafing raid made by a
single, small airplane. It did not necessarily have anything to do
with the famed and versatile de Havilland Mosquito.

$$\_\_\_\_\_\_\_\_\_\_\_\_\_\_\_\_\_\_\_\_\_\_\_ A649$$

The rocket-powered Messerschmitt Me.163 B-1a Komet, which
had a maximum speed in level flight of 559 mph, an initial climb
rate of 11,800 fpm, and an endurance of 7.5 minutes. More than
350 were produced.

_______________________________________ A650

Named by U.S. military personnel during World War II, this fictitious character flew more than 30 missions.

_______________________________________ A651

Lt. Edward "Butch" O'Hare, after whom Chicago's famous airport is named. (The identifier, ORD, stems from the airport's original name, Orchard Field.)

_______________________________________ A652

True. Silent, single-reel "shorts" were shown aboard several German airlines. Sound tracks would have gone unheard in the noisy airplanes of that era.

_______________________________________ A653

The whistle was designed and added to terrify those on the ground.

_______________________________________ A654

True. On September 9, 1942, a Yokosuka E14Y was launched from a submarine and dropped four 168-pound phosphorous bombs near Brookings, Oregon. No one was injured and damage was minimal.

_______________________________________ A655

True. General George McClellan's Union Army used manned balloons to monitor enemy troop movement.

_______________________________________ A656

The Berlin Airlift provided a flow of food (vittles) and supplies to the isolated city for more than a year (beginning in 1948).

____ A657

When dropped from low altitude, parachutes gave pilots of slow aircraft time to escape the target area before the bombs exploded.

____ A658

The machine guns of a P-51 Mustang were fed by ammunition belts that were 27 feet long. After a pilot emptied his guns on a target, he would say that he "gave 'em the whole nine yards."

____ A659

The Army's largest bombardier school was in Midland. The sooty, soft-rubber eyepieces of the bombsights left black circles around the eyes of the student bombardiers.

____ A660

False. A sortie involves a single aircraft. *Sortie* originated from the French verb, *sortir*, which means *to go out*.

____ A661

As aerial observers, they could detect camouflage based on color-blending deception more easily than crewmembers with normal vision.

____ A662

A balloon, *L'Entreperant*, was tethered by the French on June 26, 1794 during the Battle of Fleurus. From this lofty vantage point, a Colonel Coutelle observed the battlefield and directed artillery fire for nine hours.

____ A663

Axis Sally, a German broadcaster, who also was known as the Berlin Bitch.

$$\underline{\hspace{10cm}}\ A664$$

1947 and 1952, respectively.

$$\underline{\hspace{10cm}}\ A665$$

Edward "Eddie" V. Rickenbacker commanded the 94th ("Hat in the Ring") pursuit Squadron during World War I and shot down 26 German aircraft in less than six months.

$$\underline{\hspace{10cm}}\ A666$$

According to a U.S. Navy publication, the expression arose from an incident during which the wheel of a low-flying airplane struck a pedestrian on the head and crushed his top hat. For your interest and amusement, flat-hatting in German is *Flugenkerflatzmittendorfergesellshafftvereingang*.

$$\underline{\hspace{10cm}}\ A667$$

D. The 14th Voluntary Bombardment Squadron was the forerunner of the Flying Tigers, which became so named by the Chinese in December, 1941 following the Japanese attack on Pearl Harbor.

$$\underline{\hspace{10cm}}\ A668$$

True. They fought in North Africa and were named after the squadron of American pilots who were impatient with U.S. neutrality and volunteered to fight for the French prior to America's entry into World War I.

$$\underline{\hspace{10cm}}\ A669$$

They were named after the typhoons (divine winds) that saved Japan from invasion in the 1280s. Japan hoped that the modern-day *kamikazes* would be equally effective in preventing a feared invasion by the United States.

1. sleeping
2. landing with "one wing low"
3. navigating by following a railroad track
4. keeping the tail high for as long as possible during the landing roll after a wheel landing in a taildragger
5. an enemy bomber formation
6. dropping bombs
7. cockpit
8. relax, and stop talking
9. standing nervously in the commander's office while being chastised
10. jilted by a girlfriend
11. a letter to or from a girlfriend

A671

True. After being fired, the missile pulled out and unwound a lengthy wire from a drum within the aircraft. As the wire played out, the pilot could electronically and visually guide the missile to its target.

A672

Russian World War I ace Staff Captain Alexander Kazabov trailed a grappling hook on a long rope and used it to attack and destroy German aircraft.

A673

Bearcat (F8F), Cougar (F9F), Hellcat (F6F), Panther (F9F), Tiger (F11F), Tigercat (F7F), Tomcat (F14), and Wildcat (F4F). There also was the experimental Jaguar (XF10).

A674

American Airlines initiated its program in 1981.

A675

The Lockheed P-38 Lightning, which shot down more Japanese aircraft (approximately 1,800) than any other Army Air Force aircraft, including the airplane carrying Admiral Isoroku Yamamoto, the man who planned the attack on Pearl Harbor. (The Navy's Grumman F-6F Hellcat downed 5,163 Japanese aircraft.) German pilots referred to the P-38 as *der gebelschwanz teufel*, "the fork-tailed devil."

A676

Luftwaffe means *air weapon*.

A677

False. A Douglas C-47 (a military variant of the DC-3), however, was converted into a glider and renamed the XCG-17. The conversion occurred too close to the end of the war to be of value.

A678

These were worn-out (weary) B-17s and B-24s that were loaded with explosives and flown into enemy targets by remote control. (Takeoffs were made by pilots who bailed out after the aircraft had become airborne.)

A679

Dawn patrol is a military term referring to a dawn departure for the purpose of observing any shift in enemy lines or ground forces that might have occurred during the previous night.

A680

Capt. Charles A. Lindbergh (Army Air Corps) was the first military recipient. (He also was awarded the Medal of Honor.) The only civilian to receive a DFC was Amelia Earhart.

A681

Called *flechettes*, which is French for *small arrows*, the darts were dropped from aircraft with the intent of deflating airships and balloons.

A682

Introduced in 1934, the revolutionary P-26 Peashooter was Boeing's fastest fighter and for a while was America's fastest. It was called a *Peashooter* because it had a long tubular gun sight that resembled a pea shooter.

A683

Germany's Erich Hartmann (World War II, 352 victories). America's greatest ace is Richard Bong (World War II, 40 victories).

A684

The practice originated at Fort Benning, Georgia, by members of the 505th Infantry Battalion who attended the 1939 motion picture *Geronimo* the night before their first jump.

A685

Strafe comes from the German verb *strafen*, which means to punish.

CHAPTER 7
FACTS OF
FLIGHT

True or False? The purpose of small holes in the tips of wooden propellers is to allow for termites to escape or to be flung out at high rpm.

Q 687

True or False? Most lightplanes have at least one UHF receiver.

Q 688

True or False? Engine manufacturers urge pilots not to allow the manifold pressure (in inches) of naturally aspirated (non-turbocharged) engines to exceed propeller rpm (in hundreds).

Q 689

If a pilot takes off with a clogged fuel vent, he can expect
    A. fuel contamination.
    B. fuel exhaustion.
    C. fuel starvation.
    D. increased fuel flow.

Q 690

An airplane cruising at 5,000 feet MSL has a true airspeed of 200 knots, an indicated airspeed of 175 knots and a calibrated airspeed of 170 knots. The actual speed of the air flowing through the pitot tube is
    A. 000 knots.
    B. 170 knots.
    C. 175 knots.
    D. 200 knots.

Q 691

The normal mixture for a piston engine is
    A. 15 ounces of air for each ounce of fuel.
    B. 30 pounds of air for each pound of fuel.
    C. 15 quarts of air for each quart of fuel.
    D. 30 pints of air for each pint of fuel.

A pilot flying a Cessna Skylane at 10,000 feet MSL should be careful about drinking coffee from a thermos bottle because

A. coffee is a diarrhetic.

B. caffeine at altitude is debilitating.

C. the combination of coffee and supplemental oxygen induces gas.

D. opening the thermos might cause the coffee to boil and cause personal injury.

Arrange avgas, kerosene, oil, and water in order of their weight so that the lightest liquid is first and the heaviest is last.

A. kerosene, water, avgas, oil

B. oil, water, kerosene, avgas

C. water, kerosene, avgas, oil

D. avgas, kerosene, oil, water

During a normal, power-off glide at constant airspeed and sink rate, the vertical component of lift

A. is greater than aircraft weight.

B. is equal to aircraft weight.

C. is less than aircraft weight.

D. Cannot be determined.

A cargo plane filled with uncaged, live doves is in cruise flight at 7,000 feet. The birds—which are at rest on the floor—suddenly fly toward the ceiling of the airplane. This causes the airplane to

A. lose altitude and lose weight.

B. lose altitude and maintain weight.

C. maintain altitude and lose weight.

D. maintain altitude and maintain weight.

Each increase in ambient temperature of 1°F increases density altitude by approximately

A. 60 feet.
B. 120 feet.
C. 180 feet.
D. 240 feet.

Which of the following does not belong?

A. static
B. dynamic
C. viscous
D. reverted rubber

What does it mean when a wing is said to be washed out?

A. The bugs have been removed.
B. The angle of incidence has been reduced.
C. The angle of attack has been increased.
D. The wing has stalled.

An engine equipped with a non-feathering, constant-speed propeller loses oil pressure. A non-counterweighted propeller will go into _______ pitch (_______ rpm), and a counterweighted propeller will go into _______ pitch (_______ rpm).

A. low, high, high, low
B. low, high, low, high
C. high, low, high, low
D. high, low, low, high

True or False? An airplane with a thick wing is more likely to accrue wing ice (during icing conditions) than the same airplane with a thin wing.

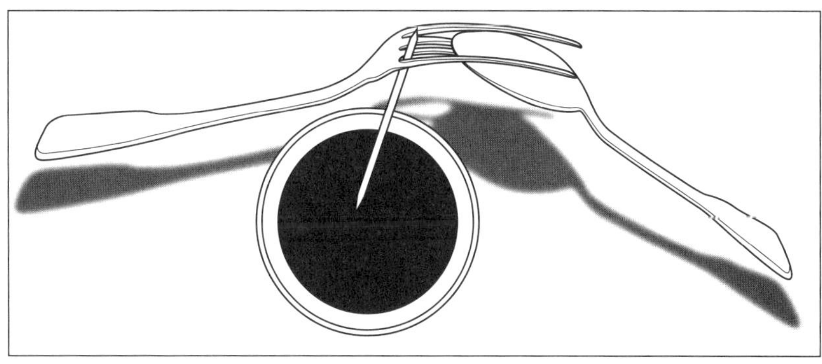

(Q701)

A spoon is forced between the tines of a fork so that the two pieces of silverware become interconnected as one (as shown). The silverware is then placed carefully on the end of a toothpick, which rests on the lip of a glass. What prevents the silverware from toppling onto the table?

True or False? The left wing of a conventional lightplane is removed and replaced with an equally long, half-inch-diameter wire that has a circular cross-section. During flight (if such a thing were possible), each wing would create the same amount of drag.

True or False? A conventional aircraft compass is called a whiskey compass because the instrument is filled with alcohol.

True or False? It is possible to safely fly an airplane equipped with a one-blade propeller.

$$Q\,705$$

Two pilots about to depart at night in an alternator-equipped airplane discover that the battery is absolutely dead. After hand-propping the engine to a start, the pilots take off and complete a routine flight. What is wrong with this scenario?

$$Q\,706$$

During a preflight runup, the pilot switches the magnetos from BOTH to LEFT and notes that there is no drop in rpm. This could mean that

- A. the right p-lead is not grounding.
- B. the left p-lead is not grounding.
- C. the right magneto had totally failed.
- D. the left magneto had totally failed.

$$Q\,707$$

What is the quaint British expression used to describe touch-and-go landings?

$$Q\,708$$

True or False? When viewed from ahead, the propellers of all U.S. piston-powered single engine airplanes turn counterclockwise, and those of British singles turn clockwise. The same can be said when looking into the inlet ducts and viewing the most forward blade sections of all U.S. and British turbofan engines.

$$Q\,709$$

Oil is used to lubricate an engine, but it also has four other purposes. Can you name them?

$$Q\,710$$

During a preflight inspection, a pilot grabs the end of the exhaust stack with a rag and discovers it to be loose or cracked. Flying this airplane could result in what great danger?

What is the quaint British expression used to describe a turn-and-bank indicator?

Batteries connected in series are used to increase _______, and batteries connected in parallel are used to increase _______.

True or False? An aircraft magneto generates AC (alternating current) electrical power.

During his preflight runup, a pilot notes that the engine is running rough on the left magneto. He suspects that the problem is caused by a bad spark plug. Without any tools other than what might be required to open the cowling, he easily determines not only if his suspicion is correct, but also which plug is misbehaving. How does he do this?

Altimeter errors are greatest when flying over mountainous terrain. Why is this so?

(Q715)

Questions

_______________________________________ **Q 716**

Which one of the world's major airports is below sea level?

_______________________________________ **Q 717**

What is the difference between an adjustable-pitch propeller and a controllable-pitch propeller?

_______________________________________ **Q 718**

What is the primary reason for shutting down an engine with the mixture control instead of by turning off the ignition?

_______________________________________ **Q 719**

An airplane has a gross takeoff weight of 2,000 pounds. A 200-pound passenger is moved 40 inches aft (from the front seat to the rear seat). How far aft does this cause the center of gravity to shift? (This problem can be solved without pencil and paper.)

_______________________________________ **Q 720**

Prior to engine start on a calm day, the pilot of a typical light airplane notes that the vertical speed indicator is displaying a 300-fpm sink rate. What is the simplest way to correct this error?

_______________________________________ **Q 721**

True or False? The terms "cowling" and "nacelle" have identical meanings.

_______________________________________ **Q 722**

What do lightning, a rifle bullet, and a whip have in common?

_______________________________________ **Q 723**

Assume that the "wings" of an airplane consist of long cylinders that spin rapidly about their own axes. If the top of the left "wing" rotates forward, and the top of the right "wing" rotates in the opposite direction, in which direction would such an airplane bank?

____ **Q 724**

True or False? The tires of high-speed airplanes are filled with helium.

____ **Q 725**

Unscramble the following letters to determine the effect of flying an airplane with an excessively aft center of gravity.
AGNITDUNLOIL BINLITISTAY

____ **Q 726**

What is the difference between a dorsal fin and a ventral fin?

____ **Q 727**

Which of the following does not belong?
  A. longeron
  B. rib
  C. stringer
  D. former

____ **Q 728**

Eight- and 12-foot-long wind socks are designed to fully extend with a wind speed of
  A. 15 knots.
  B. 20 knots.
  C. 25 knots.
  D. 30 knots.

____ **Q 729**

Which of the following is not correct when describing the stability of an airplane?
  A. positive static and positive dynamic stability
  B. positive static and negative dynamic stability
  C. negative static and positive dynamic stability
  D. negative static and negative dynamic stability

____ **Q 730**

What do the letters in the acronyms *radar* and *LORAN* represent?

____ **Q 731**

True or False? When a pressurized airplane is in cruise flight,
the amount of air entering the cabin is greater than the amount
of air leaving the aircraft.

____ **Q 732**

Why does extending the landing gear of many retractable-gear
airplanes reduce stall speed slightly?

____ **Q 733**

What is the difference between counter- and contrarotating
propellers?

____ **Q 734**

In aviation jargon, what is a truck?

____ **Q 735**

The airflow above a wing follows the cambered (curved) upper
surface (except at excessive angles of attack) because of a phenomenon known as the _______ effect.

____ **Q 736**

Why does it appear to the casual observer (standing to the side
and at some distance) that a departing Boeing 747 climbs at a
lower airspeed than a smaller jet departing at the same speed and
climb angle?

____ **Q 737**

Most pilots are aware of the hazard caused by a broken or discon-
nected P-lead (the magneto cannot be turned off). What is the
meaning of the letter *P*?

____ **Q 738**

A pilot flying a jet airplane at high altitude for long periods of time could be concerned about freezing

    A. fuel.
    B. hydraulic fluid.
    C. oil.
    D. None of the above.

____ **Q 739**

What causes a reciprocating engine to backfire?

____ **Q 740**

What is the difference between a turbocharger and a supercharger?

____ **Q 741**

Why might jogging reduce a pilot's tolerance for high Gs?

____ **Q 742**

List the four forces responsible for yawing and/or rolling an American, single-engine airplane to the left during takeoff and/or climb.

    A. _______________________
    B. _______________________
    C. _______________________
    D. _______________________

____ **Q 743**

True or False? During stabilized cruise flight on a constant heading, the advancing rotor blade of a helicopter has more airspeed and, therefore, creates more lift than the retreating blade?

Which of the following does not belong?
  A. acceleration error
  B. dip error
  C. heeling error
  D. northerly turning error

A pilot drains a fuel sample from a wing tank containing an approximately equal mixture of 80/87 octane (red dye) and 100/130 octane (blue dye) fuel. The sample will have the appearance of
  A. water.
  B. jet A fuel.
  C. 80/87 octane fuel.
  D. 100/130 octane fuel.

What is the difference between an auxiliary fuel pump and a fuel boost pump?

Which of the following is most responsible for an airplane's reaction to turbulence?
  A. aspect ratio
  B. power loading
  C. wing loading
  D. wing span

What is the difference between a radial engine and a "corn-cob" engine?

What did old-time barbershops have in common with some modern airplanes?

Lightening holes in an airplane are related to

    A. lightning.
    B. St. Elmo's fire.
    C. static discharges.
    D. weight and balance.

According to FAA's *Instrument Flying Handbook*, "If the variation between field elevation and indicated altitude is on the order of plus or minus _______ feet, the accuracy of the altimeter is questionable, and the problem should be referred to an instrument repair station."

The world's longest runway is _______ feet long and is located at _______________.

The longitudinal, lateral, and directional stability of an airplane refers, respectively, to stability about the

    A. longitudinal, lateral, and vertical axes.
    B. lateral, longitudinal, and vertical axes.
    C. vertical, longitudinal, and lateral axes.
    D. lateral, vertical, and longitudinal axes.

True or False? The decision altitude for an ILS approach to Kennedy Airport is 13,510 feet MSL.

If an aircraft engine has a fuel consumption of 20 gph, how many gallons of uncompressed air does it consume per hour?

Ignoring pilot physiology and cargo bay constraints, if a Cessna 172 is pushed downward from an orbiting space shuttle, would the flight controls become effective soon enough to prevent damage due to frictional heating and allow the aircraft to be glided safely to earth?

True or False? Flutter, which can destroy an airplane or glider, is a function of indicated airspeed.

Why is the aluminum skin of the control surfaces of some light airplanes corrugated?

A turbocharged airplane is at rest with the engine shut down on an airport with an elevation of 300 feet MSL at a time when the altimeter setting is 29.72 inches and the ambient temperature is 10°C. What should be the indicated manifold pressure of the engine?

A. 29.42 inches
B. 29.62 inches
C. 29.72 inches
D. 30.02 inches

After loading a Boeing 747, an airline cargomaster decides to move a 1,000-pound weight 50 feet forward. This causes the jumbo jet's center of gravity to move forward one inch. What is the gross weight of the airplane?

What is the difference between a de-rated engine and a flat-rated engine?

_______________________________________ **Q 762**

What is the significance of the lower limit of the green arc on a manifold-pressure gauge?

_______________________________________ **Q 763**

For those who fly in pressurized airplanes, what is the difference between an explosive and a rapid decompression?

_______________________________________ **Q 764**

British pilots call them spats; American pilots call them
_______________.

_______________________________________ **Q 765**

True or False? Your parked airplane is coated with volcanic ash. The best way to remove the ash is to blow it off with an air hose.

_______________________________________ **Q 766**

When a conventional six-cylinder aircraft engine turns at 2,400 rpm, how many combustion cycles occur per hour?

    A. 432,000
    B. 864,000
    C. 1,728,000
    D. 3,456,000

_______________________________________ **Q 767**

A .38-caliber bullet is fired through the cabin wall of a pressurized jetliner cruising at Flight Level 350. Unless the pilot makes a rapid descent, the result most likely will be

    A. no change in cabin pressure.
    B. a slow reduction in cabin pressure.
    C. a rapid decompression.
    D. an explosive decompression.

Q 768

True or False? Relatively cool air at altitude cools an overheating engine better than relatively warm air at sea level.

Q 769

What is the definition of a slow airplane?

Q 770

True or False? A turbocharged engine developing 36 inches of manifold pressure and 2,700 rpm at sea level produces 300 hp. The same manifold pressure and rpm at 20,000 feet MSL results in the same horsepower.

Q 771

The centrifugal force acting on each propeller blade of a typical general aviation airplane during takeoff is closest to

    A. 400 pounds.
    B. 4,000 pounds.
    C. 40,000 pounds.
    D. 400,000 pounds.

Q 772

True or False? A pilot stands facing a 20-knot wind that later increases to 60 knots. The force against the pilot's body increases by a factor of six.

Q 773

Estimate within 5,000 feet the absolute ceiling of the highest-flying birds.

Sensenich Propeller is the only U.S. manufacturer of certified wood propellers. These are made from

A. birch.
B. oak.
C. spruce.
D. walnut.

The term brake horsepower is used in conjunction with reciprocating engines. What is the significance of BHP?

Cessna 172s manufactured prior to and including 1986 were equipped with three fuel drains. The more modern Skyhawks (172Rs) are equipped with

A. one master fuel drain.
B. three fuel drains.
C. seven fuel drains.
D. 13 fuel drains.

Which of the following best explains why air flows faster above a conventional wing than below it?

A. Coriolis force
B. Bernoulli's principle
C. Newton's laws of motion
D. Venturi effect

Which of the following does not belong?

A. turbine engine
B. turbofan engine
C. turbojet engine
D. turboprop engine

Which of the following does not belong?

- A. A
- B. H
- C. T
- D. V

True or False? A rapid and total loss of engine oil in flight is indicated by a loss of oil pressure and confirmed by an indication of increased oil temperature.

With respect to a biplane, why doesn't the high-pressure area under the top wing interfere with and reduce the effectiveness of the low-pressure area above the bottom wing?

In terms of what is best for the engine, the least-harmful way to simulate the sudden failure of an engine developing cruise power is to

- A. close the throttle.
- B. place the mixture control to idle cutoff.
- C. turn off the magnetos.
- D. turn the fuel valve to the OFF position.

True or False? While taking off from a sea-level airport in an airplane with a normally aspirated engine and a constant-speed propeller on a standard day, the indicated manifold pressure is 29.92 inches of Hg.

What is the quaint word used by British pilots when referring to an airplane's control stick?

_______________________________ **Q 785**

Modern and sophisticated aircraft simulators have six freedoms of motion. Can you name them?

_______________________________ **Q 786**

Why do many airplanes have pull-type circuit breakers while many others have flush-type breakers?

_______________________________ **Q 787**

Explain why rain entering a pitot tube during flight neither floods the airspeed indicator nor damages the instrument's internal mechanism.

_______________________________ **Q 788**

Which of the following does not belong?
- A. interference drag
- B. form drag
- C. parasite drag
- D. skin-friction

_______________________________ **Q 789**

To dress a propeller is to
- A. file the leading edges.
- B. repaint the blades.
- C. balance it.
- D. prepare it for shipping.

_______________________________ **Q 790**

A 40-gallon fuel tank in the wing of an airplane is filled with fuel having a temperature of 60° Fahrenheit. The temperature later climbs to 90°, which causes the fuel to expand and spill overboard through the vent. The amount of spillage is 4/10ths of a
- A. cup.
- B. pint.
- C. quart.
- D. gallon.

Speaking of fuel, which of the following contains the most chemical energy per gallon?

    A. jet A (kerosene)
    B. 100LL (avgas)
    C. unleaded, regular, automobile fuel
    D. ethanol

A pilot's blood will boil in unpressurized flight at approximately

    A. 56,000 feet.
    B. 63,000 feet.
    C. 70,000 feet.
    D. 77,000 feet.

True or False? Thermal runaway is a hazard associated only with NiCad batteries.

A pilot inadvertently fails to secure a manual, plunger-type primer after priming an engine prior to starting. What is the most likely result?

True or False? Many fixed-wing airplanes are capable of inverted flight. There also is a species of bird that can fly upside-down.

A subsonic aircraft flies slower than the speed of sound, and a supersonic aircraft flies faster than Mach 1.0. What is the speed capability of a hypersonic aircraft?

____ **Q 797**

Increasing an airplane's airspeed while in level flight also
increases the speed of the air flowing past the static port(s).
Why doesn't this additional airspeed cause a reduction in
static pressure in the vicinity of the static ports and produce
erroneous altitude, airspeed, and vertical-speed (VSI) indications?

____ **Q 798**

If increasing airspeed during level flight were to cause static
ports to sense reduced ambient pressure as discussed in the
previous question,

A. the altimeter, VSI, and airspeed indicator would indicate
excessively high.

B. the altimeter, VSI, and airspeed indicator would indicate
abnormally low.

C. the altimeter and VSI would indicate excessively high, and
the airspeed indicator would indicate abnormally low.

D. the altimeter and VSI would indicate abnormally low, and
the airspeed indicator would indicate excessively high.

____ **Q 799**

Oil weighs 7.5 pounds per gallon, and avgas weighs 6.0 pounds
per gallon. How much does a gallon of water weigh?

____ **Q 800**

True or False? The pilot of a Lockheed U-2 "spyplane" is in a
normal turn at 75,000 feet MSL. He might notice that the low
wing is stalling (a stall buffet) at the same time that the high wing
is generating a Mach buffet (the result of flying in excess of the
airplane's maximum-allowable airspeed). Because each such buffet
feels the same, the pilot does not know whether to increase or
decrease airspeed to resolve the problem.

____ **Q 801**

Under what circumstances should a pilot start an engine with
the fuel valve in the Off position?

The Beech Bonanza Model 35 has a V-tail configuration with rud-
dervators, control surfaces that combine the function of rudder
and elevator. What are elevons, and on what kind of aircraft are
they most frequently found?

Carburetor ice can form when the outside air temperature is
substantially warmer than 32°F. What causes the temperature
to drop so dramatically in the throat of a carburetor?

The most northerly civilian airport in the world is in
  A. Canada.
  B. Greenland.
  C. Norway.
  D. Russia.

True or False? The effect of magnetic dip is offset somewhat by
the addition of a small weight to the south pole of aircraft compasses destined for use in the Northern Hemisphere. Similarly,
a small weight is added to the north pole of compasses intended
for use in the Southern Hemisphere.

True or False? All controlled airports in the United States have
at least one hard-surface runway.

A pilot owns a 120-knot airplane powered by a 200-hp engine. He decides to modify the airplane with an engine that theoretically will double cruise performance. Everything else being equal, the new engine must produce approximately _______ times as much power.

A. 2 (400 hp)
B. 4 (800 hp)
C. 6 (1200 hp)
D. 8 (1600 hp)

Which of the following does not belong?

A. differential
B. Fowler
C. Frise
D. slotted

A pilot makes a forced landing in the boondocks and the aircraft is destroyed. He knows that parts of the wreckage can be used to enhance his likelihood of survival until a rescue team arrives. How can he make best use of the tires?

Why can overpriming an engine cause a loss of compression?

What is the relationship between dihedral, anhedral, and cathedral?

Cabin heat in most single-engine airplanes is obtained by using hot exhaust gasses to heat ambient air flowing through a muff. Considering that many twins have cabins that are no more voluminous than many of these singles, why are they equipped with expensive, gasoline-fired heaters to achieve the same result?

Which of the following does not belong?
- A. decalage
- B. open cockpit
- C. sesquiplane
- D. stagger

True or False? An airplane in a constant-rate, coordinated turn while maintaining a constant altitude at a fixed airspeed is not accelerating.

True or False? A pilot can experience blackout when exposed to sufficient positive Gs. The opposite of a blackout occurs when exposed to sufficient negative Gs and is called a red-out.

True or False? Afterburners produce a burst of additional jet thrust on some aircraft. This is accomplished by igniting unused fuel and oxygen that escape from the combustion chamber(s) and flow through the turbine(s) and into the tailpipe of such an engine.

What is an airspeed bomb?

____ **Q 818**

The total wing area of a multiengine airplane with wing-mounted engines _____________ that portion of each wing occupied by the nacelles and _____________ that area of the wing projected through the fuselage.

    A. includes, includes
    B. includes, excludes
    C. excludes, excludes
    D. excludes, includes

____ **Q 819**

A non-cantilevered wing has at least one primary strut to provide support. Small or auxiliary struts often are used in conjunction with and to support the main wing struts. These are called

_____________________.

____ **Q 820**

True or False? Modern aviation transceivers operating in the VHF communications band (118.000-136.975 MHz) utilize frequency modulation (FM).

____ **Q 821**

True or False? When the fuselage of an airplane is stretched and the fuselage plug is added forward of the quarter-chord of the wing (such as when Beech stretched the F-33A into the A-36), this has the effect of increasing yaw stability.

____ **Q 822**

A pilot shuts down an engine, but it continues to operate in a somewhat irregular manner for some time after the ignition is turned off. In an automobile, this is caused dieseling. When occurring in an airplane, it is called _______________ and is caused by _______________.

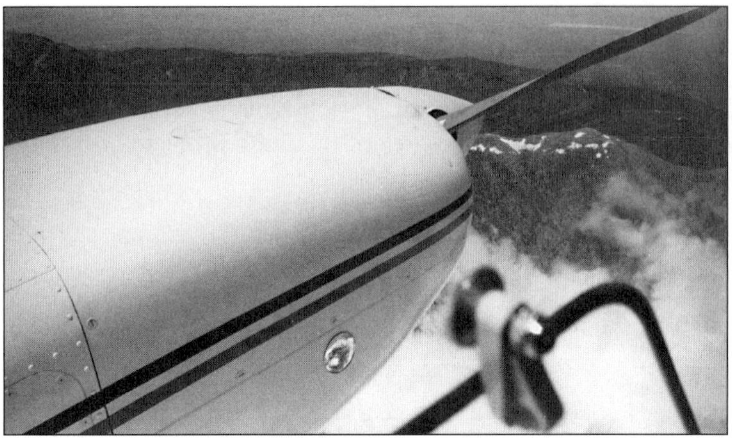

Feathered propeller of a Piper Seminole. (Q823)

____ Q 823

The primary purpose of feathering a propeller is to reduce drag and improve engine-out performance. What are two other important reasons for feathering?

____ Q 824

British pilots refer to high- and low-rpm propeller settings as _______ and _______ pitch, respectively.

____ Q 825

Many aircraft, including early model Beech Bonanzas, were equipped with wobble pumps. What is a wobble pump?

____ Q 826

Why does moving the center of gravity forward make an airplane more stable in pitch and yaw?

____ Q 827

The North American XB-70 Valkyrie supersonic bomber's six engines were clustered in a single horizontal row beneath the aircraft. This was colloquially known as a *six-pack* (of engines). With what six-pack (having nothing to do with engines) are most general aviation aircraft equipped?

____ **Q 828**

What is the most commonly accepted method used to clear a
flooded engine?

____ **Q 829**

True or False? FAA's *Operation Lights On* encourages pilots to
turn on their landing lights to enhance the see-and-avoid concept
of collision avoidance when operating near an airport. With
respect to extending the longevity of light bulbs, it is better to
leave the lights on for the entire flight than to turn them on and
off at the beginning and end of a flight.

____ **Q 830**

Why does moving the center of gravity aft cause stall speed to
decrease (and vice versa)?

____ **Q 831**

A pilot is about to depart Honolulu International Airport at local
noon on or about either May 27 or July 16. During his preflight
inspection, what unusual and strange phenomenon might he
notice that never occurs in any of the other 49 states?

____ **Q 832**

On many lightplanes, there is a control-surface extension that
protrudes forward of the rudder or elevator hinge line. What is
the purpose of such an extension?

____ **Q 833**

The maximum outside-air temperature at which carburetor ice
can occur is approximately

    A. 107°F (41.7°C).
    B. 97°F (36.1°C).
    C. 87°F (30.6°C).
    D. 77°F (25°C).

The *lomcovák* (an aerobatic maneuver) is a Czechoslovakian word derived from the name or word that means or represents

A. a large, stiff drink.
B. a migraine headache.
C. a town in Czechoslovakia.
D. its creator's mother-in-law.

If *topping off* means to fill the fuel tanks, what does *topping up* mean?

A pilot installs a venturi tube on his airplane to power his gyroscopic flight instruments. The cross-sectional area of the throat of the tube is half of the cross-sectional area of the entrance. During cruise flight at 100 knots at sea level, the airspeed inside the throat of the venturi is doubled, and static pressure is reduced by approximately

A. 1.5 inches Hg.
B. 4.5 inches Hg.
C. 7.5 inches Hg.
D. 15.0 inches Hg.

True or False? A propeller brake was used on some airplanes to mechanically arrest the rotation of a propeller following engine failure.

With the exception of Rogers (dry) Lake at Edwards Air Force Base in southern California, which is several miles long, what is the longest paved runway in the United States, and how long is it?

While flying an aircraft with a fuel-injected engine, a pilot notices that the head temperature of one cylinder is significantly hotter than the others. What is the most likely cause of this overheating?

When an American flies in England, he is likely to hear some unusual nomenclature. What expressions might a British pilot use when referring to

A. a wet-cell battery?
B. fuel?
C. a horizontal stabilizer?
D. landing gear?
E. leveling off (at the end of a climb or descent)?
F. a propeller?
G. a wing?

To increase the likelihood that cockpit voice recorders (CVRs) and flight data recorders (FDRs) can survive an accident, they are built to withstand _____ Gs (in 6.5 milliseconds) and saltwater pressure equivalent to that found at an ocean depth of _______ feet.

A. 1,700, 10,000
B. 1,700, 20,000
C. 3,400, 10,000
D. 3,400, 20,000

Author performing a left forward slip in a
Bellanca Citabria 7GCAA. (Q842)

____ Q 842

Why can taildraggers typically be made to slip more steeply
(and have steeper descent angles) than aircraft with tricycle
landing gear?

____ Q 843

What would cause an airspeed indicator to behave like an altimeter? In other words, indicated airspeed would increase with an
increase in altitude and decrease with a decrease in altitude,
irrespective of pitch and/or power changes.

Which of the following describe(s) the changes that occur to a flow of subsonic air as it enters a venturi tube and approaches the throat (narrowest cross-section)?

A. airspeed decreases
B. airspeed increases
C. density decreases
D. density increases
E. pressure decreases
F. pressure increases

Which of the following does not belong?

A. Fowler
B. Friese
C. slotted
D. split
E. zap

Many altimeters incorporate a window on the face of the instrument through which is displayed a "flag" of diagonal, black and white (or black and yellow) stripes. What is the significance of this flag?

Why is that wind-direction (runway-in-use) indicator found at many airports called a *tetrahedron*?

Everything else being equal, a properly inflated tire wears evenly across its tread. The tread wears more toward its center when the tire is _______ and more toward its shoulders when _______.

A. overinflated, overinflated
B. overinflated, underinflated
C. underinflated, overinflated
D. underinflated, underinflated

True or False? Some jet fighters have roll rates in excess of 720 degrees per second, but a barn swallow can roll in excess of 5,000 degrees per second and pivot its head nearly 360 degrees during flight.

Vertigo and spatial disorientation can result in the loss of aircraft control. What is the difference between these phenomena?

What is the difference between spoilers and speed brakes?

True or False? With respect to horizontally opposed Teledyne Continental and Avco Lycoming engines, the number-one cylinder is the rearmost cylinder on the right.

An attitude indicator shows changes in pitch and roll because of the way in which a spinning gyroscope remains fixed in space. When then doesn't the conventional artificial horizon indicate an inverted attitude following a nonstop flight from the North to the South Pole?

What is the world's fastest bird, and what is its top speed?

True or False? The African eagle can swoop at more than 100 mph and aerodynamically brake to a halt in only 20 feet.

False. The small holes allow moisture to escape.

True. A transponder receives interrogation at 1,030 MHz and replies on 1,090 MHz. Both are in the UHF band of 300 to 3,000 MHz.

False. If this wive's tale were true, every low-altitude takeoff would violate such a recommendation. The most reliable advice for operating an engine is found in the Pilot's Operating Handbook.

C. A clogged fuel vent prevents air from entering the tank to replace consumed fuel. This results in a reduction of air pressure in the tank. Such a "suction" eventually prevents fuel flow from the tank to the engine and results in fuel starvation. The engine fails even through there is fuel in the tank. For those who confuse the terms *fuel starvation* and *fuel exhaustion*, just remember that "an engine can be starved of fuel even if the fuel supply has not been exhausted."

A. Air does not flow through a pitot tube. Where would it go? Movement through the atmosphere causes pressure to increase in the pitot system, and it is this pressure that results in indicated airspeed.

A. The correct stoichiometric mixture (or chemical combination) for a piston engine is 15 parts of air to one part of fuel, by weight, not by volume.

_______________________________________ A_{692}

D. The boiling point of water decreases with altitude. If the temperature of the coffee in the thermos is greater than the boiling
point of water at some given altitude, opening the thermos
would allow the coffee to suddenly boil and scald anyone
sprayed by the unexpected eruption.

_______________________________________ A_{693}

D. The only thing that a pilot must know to answer this question
is that water weighs more than avgas (which it obviously does
or it would not settle to the bottom of a fuel tank or strainer).
D is the only answer that indicates this relationship between
avgas and water.

_______________________________________ A_{694}

B. In stabilized flight, lift equals weight, and thrust equals drag.
If weight were greater than lift, the aircraft would accelerate
downwards. In other words, the sink rate would steadily
increase. If lift were greater than weight, acceleration would
occur in an upward direction.

_______________________________________ A_{695}

B. Doves flying toward the ceiling would cause the airplane's center of gravity to rise. Unless the forces acting upon the airplane
change, it is the CG that maintains a constant altitude, not the
airplane. Therefore, the airplane must descend slightly so that
the overall CG does not change altitude. Aircraft weight would
not change because none is removed from the aircraft. (The air
from the birds' wings would beat upon the floor with a total
force equal to the weight of the birds.)

_______________________________________ A696

A. Density altitude can be approximated by adding 60 feet to the elevation of an airport for each 1°F that the ambient temperature is greater than the standard temperature for that elevation.

_______________________________________ A697

A. There are three types of hydroplaning: dynamic, viscous, and reverted rubber. There is no such thing as static hydroplaning.

_______________________________________ A698

B. Airplane designers washout (or twist) some wings so that the wing tips have smaller angles of incidence than the inboard wing sections. Washout is used to improve the stall characteristics of a wing.

_______________________________________ A699

A. Losing oil pressure causes a non-counterweighted propeller to go into low pitch (high rpm) and a counterweighted propeller to go into high pitch (low rpm).

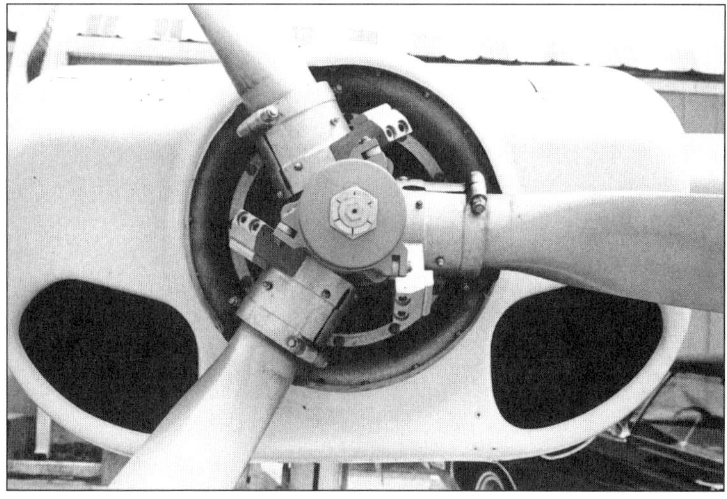

Constant-speed propeller.   (A699)

Answers

____ A700

False. Thin, small-radius objects accrue ice more readily than do objects with a large radius. This is why antennas, windshield wipers, and the like usually accrue ice before the airframe does.

____ A701

The silverware balances on the toothpick because the moments on each side of the fulcrum are equal. The center of gravity is at the fulcrum, the point at which the toothpick rests on the glass. The handles of the silverware (which are on one side of the fulcrum) are exactly balanced by the joined ends (which are on the other side of the fulcrum).

____ A702

True. A circular wire creates much more drag than many realize and is one reason why old airplanes with numerous flying wires and big engines could not go very fast.

____ A703

False. Alcohol was never used in aircraft compasses but was used in maritime service, which is where the term originated. The "wet" compass in an aircraft is filled with highly distilled kerosene.

____ A704

True. There have been a number of experimental one-blade propellers. A counterweight provided balance, but because all thrust was generated on one side of the propeller disc, the crankshaft and its bearings were seriously strained.

$$A705$$

An alternator cannot function without a slight current flow from the battery. Therefore, if the battery is totally discharged, the alternator will not operate. Required aircraft lights would be inoperative.

$$A706$$

Two of the four answers are possible. Either the engine continues to run on both magnetos because of (A), or it has been running only on the left magneto because of (C).

$$A707$$

Circuits and bumps.

$$A708$$

True with respect to propellers but false with respect to turbofan engines.

$$A709$$

Oil also is used to cool the engine, clean it, cushion the movement of moving parts, and protect against corrosion.

$$A710$$

Hot exhaust could enter the cowling and cause an engine fire.

$$A711$$

A bat and ball.

____ A712

Batteries in series increase voltage; batteries in parallel increase amperage.

____ A713

True. Magnetos are independent of the electrical systems of light airplanes, which typically produce DC (direct current).

____ A714

The pilot returns to his parking place while operating the engine only on the malfunctioning magneto, which gives the cylinder with the bad plug a chance to cool off. After running the engine for several minutes in this manner, he shuts down the engine and quickly opens the cowling. He then carefully feels the top (rocker arm cover) of each cylinder with the palm of a hand. (Another way is to touch or spit on each exhaust riser.) If one cylinder feels relatively cool to the touch, the pilot has verified that this is the cylinder with the bad plug. This technique can save considerable time troubleshooting such a problem.

____ A715

Wind flowing over and around mountains results in a venturi effect much like air flowing over a wing. This causes wind speed to increase and (static) pressure to decrease, which can cause indicated altitude to be significantly greater than true altitude.

____ A716

Amsterdam's Schiphol International Airport is 11 feet below sea level.

Controllable-pitch propeller.

Adjustable-pitch propeller. (A717)

$$A717$$

The blade angles of an adjustable-pitch propeller can be changed only on the ground (by loosening the clamps with a special tool and manually adjusting the blade angles); the blade angles of a controllable-pitch propeller can be adjusted from within the cockpit during flight. In each case, blade angles remain subsequently fixed irrespective of power and/or airspeed changes. Unlike a constant-speed propeller, power and/or airspeed changes do result in a corresponding change in rpm.

$$\overline{}\ \text{A718}$$

Using the mixture control allows fuel in the cylinders to be consumed during shutdown. If the engine is shut down by turning off the magnetos, a hot spot in a cylinder can ignite remaining fuel after shutdown and endanger someone near the propeller.

$$\overline{}\ \text{A719}$$

Four inches. The CG shift is found by multiplying the weight moved by the distance it is moved and dividing the result by the gross weight of the airplane. (200 times 40 divided by 2,000 = 4)

$$\overline{}\ \text{A720}$$

Virtually every conventional VSI has a small set screw on the front of the instrument that is used to adjust the indicator. Only a certified repairman is allowed to do this. *<wink>*

$$\overline{}\ \text{A721}$$

False. A cowling usually is part of the fuselage (such as on a single-engine airplane), whereas a nacelle (such as on a conventional twin) is not.

$$\overline{}\ \text{A722}$$

Each creates a sonic boom.

$$\overline{}\ \text{A723}$$

The airplane would bank left. This is called the Magnus effect and is the same phenomenon that explains the behavior of a pitched and spinning baseball.

$$\overline{}\ \text{A724}$$

False. They are filled with nitrogen, which does not support combustion that can result from excessively hot tires.

LONGITUDINAL INSTABILITY

A dorsal fin is installed on the top of a fuselage, and a ventral fin is located on the bottom. (Memory aid: d comes before v in the alphabet). The fin on the top of a shark, therefore, is a dorsal fin, not a ventral fin.

B. A rib is a structural member of a wing. The others are parts of a fuselage.

A. The sock also must be capable of withstanding a wind speed of at least 75 knots.

C and D. If an airplane has negative static stability, it cannot have either positive or negative dynamic stability.

RAdio Detecting And Ranging, and LOng RAnge Navigation.

False. If that were so, the cabin would inflate like a balloon until the pressure vessel (cabin) eventually ruptures.

The air resistance of the landing gear forces a slight amount of air that would otherwise pass under the wing to flow over the wing, which reduces stall speed.

_____________________________________ A733

Contrarotating propellers are mounted on the same axis of rotation but turn in opposite directions. Counter-rotating propellers also turn in opposite directions but are mounted on different engines.

_____________________________________ A734

A truck is a landing gear leg that usually has four wheels (or more) arranged like those on an automobile (or truck).

_____________________________________ A735

Coanda. This behavior of fluid flow is named after the French engineer and inventor, Henri Marie Coanda.

_____________________________________ A736

Large airplanes appear to move slower because it takes longer for them to travel a distance equal to their own length, which is the subconscious reference that an observer uses to estimate speed.

_____________________________________ A737

The letter *P* represents the primary lead (or circuit) of the magneto.

_____________________________________ A738

A. This is why pilots of the Boeing 747-100, for example, must maintain an indicated fuel temperature that is at least 3°C above the freezing point of the fuel being used.

_____________________________________ A739

The fuel-air mixture in the induction system is ignited by gases that are still burning in a cylinder when its intake valve opens.

Compressor and turbine portions of a turbocharger.   (A740)

______________________________________ A740

A turbocharger is exhaust driven, and a supercharger is mechanically driven.

______________________________________ A741

Jogging tends to lower blood pressure, which can reduce high-G tolerance.

______________________________________ A742

A. P-effect (or P-factor)

B. gyroscopic precession

C. torque

D. swirling slipstream

______________________________________ A743

False. If this were so, the rotor disk would be unbalanced, and the excess lift on the side of the advancing blade would cause the helicopter to roll forcefully toward the retreating rotor blade.

_________________________________ A744

B. The other choices are errors caused by magnetic dip, which
   is not an error. Instead, dip is the vertical component of the
   Earth's magnetic field at any given location.

_________________________________ A745

A. The red and blue dyes can neutralize each other and give the
   appearance of a clear liquid.

_________________________________ A746

Although the terms are frequently misused and interchanged,
an "aux" pump is used for starting, for takeoff, in case of engine-
driven pump failure, and to pressurize fuel to the engine pump
to prevent vapor lock at altitude. A boost pump is in a fuel tank
and provides a head of pressure to the main pump.

_________________________________ A747

C. Aircraft reaction (acceleration) to a vertical gust decreases as
   the load carried by each square foot of wing increases.

_________________________________ A748

A "corn-cob" engine is a radial engine with two or more banks
of cylinders that — when uncowled — resembles a cob of corn.

_________________________________ A749

Barber poles. On jet airplanes, these are maximum-airspeed point-
ers (variable redlines) built into airspeed indicators. They provide
the maximum-allowable airspeed, which varies with altitude.

_________________________________ A750

D. These holes are made in aircraft parts to reduce or _lighten_ aircraft empty weight. Metal is removed in locations where the
   holes would not cause any decrease in structural integrity.

$$A751$$

75 feet. The Federal Aviation Regulations, however, apparently do not specify a maximum-allowable error other than at the time of a required static-system check.

$$A752$$

The runway at Edwards Air Force Base in southern California is 24,852 feet long and is an occasional landing site for the space shuttle. (Part of the runway length consists of the natural surface of Rogers [dry] Lake.)

$$A753$$

B. Longitudinal, lateral, and directional stability are synonymous with pitch, roll, and yaw stability, respectively.

$$A754$$

True. *This* Kennedy Airport serves La Paz, Bolivia, which has an elevation of 13,310 feet MSL and should not be confused with the John F. Kennedy International Airport serving New York City.

$$A755$$

160,000 gallons. Reciprocating engines consume 8,000 parts of air (by volume) for each part of fuel.

$$A756$$

No. The extraordinary true airspeed (even at an indicated airspeed of only 50 knots) would result in damaging heat, compression, and destructive Mach airspeed effects. Not much of the structure would make it to earth.

$$A757$$

False. Flutter is a function of true airspeed, which is why V_{NE} must be reduced in some aircraft at high altitude.

$$\overline{\hspace{10cm}}\ A758$$

Corrugation increases the stiffness of a control surface while requiring a minimum of internal structure, which can reduce weight and the cost of construction.

$$\overline{\hspace{10cm}}\ A759$$

A. When the engine is shut down, a manifold pressure gauge acts as a barometer and indicates ambient atmospheric pressure, which — in this case — is equal to 29.72 inches minus 0.30 inches, or 29.42 inches. The temperature is irrelevant.

$$\overline{\hspace{10cm}}\ A760$$

600,000 pounds. The formula is the same for all airplanes: the distance a weight is moved is equal to aircraft gross weight times the distance of CG movement divided by the weight of the item(s) shifted. To solve the problem, ensure that both distances are expressed in common units (inches or feet).

$$\overline{\hspace{10cm}}\ A761$$

A de-rated engine is typically a piston engine that is limited by the manufacturer to develop less power than the power for which it was originally designed. A flat-rated engine is typically a turbo-prop engine that can develop more power (at low altitude) than the pilot is permitted to use.

$$\overline{\hspace{10cm}}\ A762$$

When manifold pressure is "in the green," the engine usually is driving the propeller, but when MP is "below the green," the propeller usually is driving the engine (windmilling).

____ A763

Explosive decompression is a change in cabin pressure that occurs faster than the lungs can decompress, which can be injurious. During rapid decompression, the lungs can decompress faster than the cabin.

____ A764

Wheel pants.

____ A765

False. According to the Alaska Airmen's Association, the best way to remove ash is to flood it off with water.

____ A766

A. Each piston/cylinder of a four-cycle engine undergoes a complete cycle every two revolutions of the crankshaft. 3 cycles/revolution x 2,400 rpm x 60 minutes per hour = 432,000 cycles/hour.

____ A767

A. Cabin air normally escapes through the outflow valve(s) in much greater volume. To compensate for the small amount of air escaping through the bullet hole, the outflow valve(s) would close slightly to prevent a pressure loss (James Bond movies notwithstanding).

____ A768

False. Dense air—even when relatively warm—carries away engine heat more efficiently than air that is less dense and relatively cool.

____ A769

A slow airplane is one that receives a bird strike from behind.

A770

False. The compressed air from the turbocharger results in a much higher induction temperature, which usually results in less horsepower (especially if the engine is not fitted with an intercooler).

A771

C. This should inspire pilots to thoroughly inspect their propellers before flight.

A772

False. Wind force increases in proportion to the square of the increase in speed. Similarly, parasite drag rises in proportion to the square of an increase in indicated airspeed (everything else being equal). Tripling airspeed (or wind speed) increases drag by a factor of nine.

A773

Whooper swans have been seen at Flight Level 290 (29,000 feet MSL), and the Alpine chough has been observed above Mt. Everest (29,028 feet MSL). Also, the highest flying bird is the Ruppell's vulture, one of which collided with a Boeing jetliner over the Ivory Coast in Africa at Flight Level 410. None of these birds, however, were squawking the appropriate transponder code.

A774

A. Sensenich used to make butcher blocks and table tops from the remaining propeller scraps. Hartzell Propeller used walnut and then birch for its wood propellers, which are no longer manufactured.

A775

Brake horsepower is the same as shaft horsepower. In aviation, each represents the amount of horsepower delivered by the crankshaft to the propeller.

$$A776$$

D. There are five under each wing and three under the nose.

$$A777$$

D. Although Bernoulli's principle explains why pressure above the wing is reduced, this occurs as a result of the increase in speed caused by Venturi effect.

$$A778$$

A. A turbine engine can be either a turbofan, turbojet, or turbo-prop engine.

$$A779$$

A. There are H-tails (such as on the Twin Beech), T-tails, and V-tails, but there are not yet any A-tails.

$$A780$$

False. The indicated oil temperature might actually decrease because oil no longer flows past the temperature-sensing probe in the engine.

$$A781$$

According to popular wisdom, such interference does occur but is insignificant when the distance between the two wings (gap) is greater than the longest chord of either wing.

$$A782$$

B. Also, the throttle should be kept open. This allows manifold pressure to remain high and keeps the combustion chambers filled with shock-absorbing air. This, however, is not necessarily the safest way to simulate an engine failure.

A783

False. Manifold pressure will be less because of pressure losses caused by restrictions to the flow of induction air such as filters, bends in the plumbing, throttle valves, and so forth.

A784

Pole, as in "pole and pedals" instead of "stick and rudder."

A785

Pitch, roll, and yaw; and lateral (sideways), vertical, and fore-and-aft movements. The latter three movements also are known as sway, heave, and surge, respectively.

A786

The less-expensive, flush-type circuit breakers are installed primarily to reduce aircraft manufacturing costs.

A787

Air and rain do not flow through the pitot tube to the airspeed indicator. Instead, ram air compresses ambient air already in the pneumatic plumbing to increase pressure measured by the instrument.

A788

C. The other three items are forms of parasite drag.

A789

A. The purpose of such filing (or dressing) is to remove stress points (risers) and should be performed only by a certified mechanic.

A790

D. According to Exxon, a 40-gallon tank will lose almost a half-gallon from expansion. The greater the temperature rise, the greater is the loss.

A791

A. In terms of chemical energy, mogas is next, 100LL is next to last, and ethanol contains the least chemical energy.

A792

B. Also, a given volume of air taken from sea level to this altitude would expand to 14.7 times its original size.

A793

False. Under certain high-load, high-temperature conditions, a lead-acid battery can seriously overcharge and explode.

A794

If the plunger works its way out to the open position during flight, fuel will be drawn through the primer and into the engine. This most likely will result in such an excessively rich mixture that the engine will lose power or fail entirely. Always verify that the plunger is locked in the closed position prior to takeoff.

A795

True. Because of its angled wing structure, the hummingbird can fly inverted for short periods.

A796

A hypersonic aircraft can fly at five times the speed of sound (Mach 5.0).

_______________________________________ **A**797

Even though the airflow (or streamlines) passing by the static ports do increase in velocity, the velocity of the streamlines ahead of the static ports is the same as the velocity of the streamlines passing by the ports, i.e., velocity does not change along the length of the streamlines in the vicinity of the static ports. According to Bernoulli's principle, a change in pressure only results from a local change in velocity (as occurs to streamlines flowing over a cambered wing).

_______________________________________ **A**798

A. Reducing static pressure causes the altimeter and VSI to behave as if the airplane were climbing. Reducing static pressure *without* a corresponding reduction in static pressure entering the pitot tube causes indicated airspeed to increase beyond what would normally be expected.

_______________________________________ **A**799

8.3 pounds per gallon. This means that water is heavier than oil and much heavier than avgas, which is why water settles so rapidly to the bottom of a fuel tank or fuel sampler.

_______________________________________ **A**800

True. The pilot is operating in "coffin corner," the upper limit of the airplane's performance envelope. The solution is to roll out of the turn to increase the speed of the inside wing and decrease the speed of the outside wing. If the buffet continues, then both wings are flying at either too low or too high an airspeed. Hopefully, the pilot can determine which is the case and make the appropriate airspeed adjustment. Otherwise, the condition can be exacerbated and result in loss of control.

$$A 801$$

When propping the engine without anyone in the cockpit. There should be enough time to get in the airplane and turn on the fuel before the engine fails. This reduces the possibility of a runway airplane. Ensure that the aircraft is tied down and chocked before starting.

$$A 802$$

Elevons are control surfaces that combine the functions of ailerons and elevators. They are most commonly found on delta-wing airplanes that do not have conventional tails.

$$A 803$$

Fuel sprayed into the throat of a carburetor evaporates, a process that consumes a substantial amount of heat. This heat is taken from the air passing through the throat and lowers air temperature there substantially. This is the same reason that your wet skin feels cold after climbing out of a swimming pool; bodily heat is used to evaporate the water. Also, adiabatic cooling is caused by expansion of air in the throat.

$$A 804$$

C. The most northerly is Svalbard Airport, which is on a Norwegian island due north of mainland Norway. At 78 degrees 14 minutes N, it is only 706 NM from the true North Pole.

$$A 805$$

True. At least one compass manufacturer, the Airpath Instrument Company, does this to reduce compass errors caused by magnetic dip. The amount of dip offset of a specific compass can be determined by observing the extent to which a compass card is tilted during straight-and-level flight.

False. At least one airport does not. Lake Hood Seaplane Base in Anchorage, Alaska consists of only freshwater landing lanes. (The FAA defines a seaplane base as an airport.)

D. Airspeed for a given airframe varies approximately with the cube root of the horsepower increase. Doubling speed, therefore, requires eight times as much power. (The cube root of 8 is 2.)

B. Fowler refers to a type of wing flap. The other three are types of ailerons. (Flaps that do not operate in unison are called flaperons; they are not called differential flaps.)

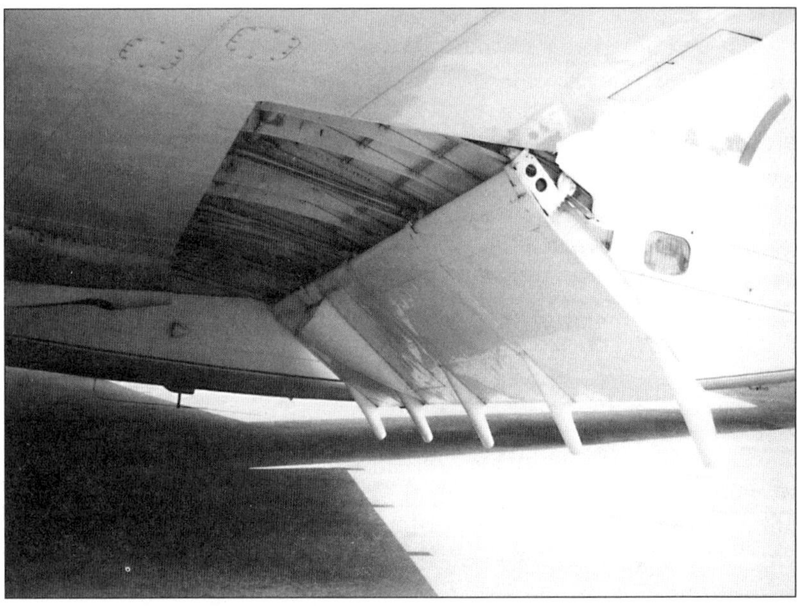

The Fowler flap on this Lockheed Lodestar first extends aft to increase wing area and then deflects downward to increase wing camber. (A808)

$$\underline{\hspace{7cm}} A809$$

Burning tires generate a substantial amount of black smoke, which increases the likelihood of being located by search-and-rescue personnel.

$$\underline{\hspace{7cm}} A810$$

The excess raw fuel washes oil from the cylinder walls and dissolves the oil seal normally formed between the piston rings and the walls.

$$\underline{\hspace{7cm}} A811$$

Dihedral can be either positive or negative. Positive dihedral is called anhedral, and negative dihedral is called cathedral.

$$\underline{\hspace{7cm}} A812$$

Ducting heated air for such a lengthy distance (from the nacelle of a wing-mounted engine to the cabin) usually results in excessive and unacceptable heat loss.

$$\underline{\hspace{7cm}} A813$$

B. The other three items relate exclusively to biplanes. Decalage is the difference in the angles of incidence between the two wings. A sesquiplane is a biplane on which one wing has less than half the area of the other (such as the Nieuport 27 of World War I).

$$\underline{\hspace{7cm}} A814$$

False. A turning airplane is in a state of acceleration caused by the horizontal component of wing lift (the turning force). Otherwise, it would continue in a straight line. It can be said that an airplane is in accelerated flight whenever its load factor is other than +1 G.

Answers

$$A 815$$

True. A blackout is caused by blood draining from the brain. The opposite occurs during negative Gs and is called red-out because excessive blood is forced to the brain; the pilot really does see red.

$$A 816$$

False. Fuel is injected into the tailpipe and burns because of unused oxygen in the exhaust. (Less than half of the available oxygen is consumed during normal combustion.) Afterburners are called reheaters in Great Britain.

$$A 817$$

An airspeed bomb is a pitot-static source that trails via a long tube below a test aircraft. It moves through undisturbed air beneath the aircraft to determine the accuracy of the installed airspeed indicator.

$$A 818$$

A. With respect to the wing projection through the fuselage, the wing area of a single-engine airplane is determined similarly.

$$A 819$$

Jury struts. Those used to connect the wings of a biplane are called interplane struts.

$$A 820$$

False. They utilize amplitude modulation (AM).

$$A 821$$

False. Everything else being equal, adding vertical surface area forward of the wing decreases yaw stability. Stretching the fuselage aft of the wing's quarter-chord moves the stabilizers farther aft and increases yaw stability.

____ **A**822

Preignition. It is caused by carbon deposits or other hot spots in one or more cylinders that can ignite fuel in the combustion chambers after the magnetos have been turned off.

____ **A**823

Feathering prevents further internal engine damage that might be caused by a windmilling propeller. It also prevents engine-driven fuel, oil, and hydraulic pumps from pumping fluids that could cause or sustain an engine fire.

____ **A**824

Fine (low pitch, high rpm) and *coarse* (high pitch, low rpm).

____ **A**825

A wobble pump is a fuel or hydraulic pump that is manually operated with a handle (like an old-fashioned water pump). If the engine-driven fuel pump of a 1947 Beech Bonanza were to fail, for example, the pilot would have to hand pump fuel to the engine until safely on the ground.

____ **A**826

Moving the CG forward increases the moment arm (leverage) of the horizontal and vertical stabilizers, which makes them more effective. An aft CG reduces pitch and yaw stability.

____ **A**827

The standard matrix of six flight instruments is similarly called a *six-pack*. These instruments consist of the airspeed indicator, attitude indicator, and altimeter (left to right, top row); and the turn-and-bank indicator (or turn coordinator), heading indicator, and VSI (left to right, bottom row).

_______________________________________ A 828

Crank the engine with the throttle fully open, the mixture control in idle-cutoff, and the fuel pump off (if installed) until the engine fires. Then increase the mixture and reduce power.

_______________________________________ A 829

True. According to General Electric, cycling light bulbs on and off is much more destructive than leaving them on (unless prohibited by the pilot's operating handbook).

_______________________________________ A 830

When the CG moves aft, the horizontal tail surfaces do not have to work as hard (produce as much downward lift) to keep the tail down. When downward lift on the tail is reduced, the wing does not have to create as much upward (and offsetting) lift. This reduces wing loading and stall speed.

_______________________________________ A 831

Hawaii is the only state south of the Tropic of Cancer. At Honolulu, the sun on these dates passes directly overhead at local noon (when it crosses Honolulu's meridian). As a result, vertical objects such as vertical fins, poles, and so forth cast no shadows.

_______________________________________ A 832

These are balanced control surfaces. The extensions are sometimes filled with lead and reduce the control force required by the pilot to move these surfaces. Such extensions also reduce the possibility of flutter.

_______________________________________ A 833

A. *Serious* icing can occur at any power setting when the ambient temperature is between 23°F and 62°F. Carburetor ice is possible only when the humidity is at least 18 percent.

A834

A. Some suggest that the result of consuming a *lomcovák* is similar to what it feels like to perform one, or that such a drink might be required before having enough courage to perform one for the first time.

A835

Topping up means to fill a partially inflated airship or balloon with gas (helium).

A836

A. So sayeth Bernoulli's theorem. A vacuum pump, however, is much more effective and typically provides anywhere from two to four times as much pressure reduction.

A837

True. They were used before the popular acceptance of constant-speed, full-feathering propellers to eliminate the drag caused by windmilling and possibly prevent further engine damage.

A838

The longest public runway is 16R/34L at Denver International Airport, which is 16,000 feet long. Unofficially, the longest paved runway in the world is 14/32 at the top-secret Groom Lake facility in "Area 51" (Nevada). It reportedly is 27,128 feet long.

A839

A partially clogged fuel injector is most likely causing the misbehaving cylinder to operate with an excessively lean fuel-air mixture.

—————————————————————— A840

A. *Accumulator*

B. *Petrol*

C. *tail plane*

D. *alighting gear* (or *undercarriage*)

E. *flattening out*

F. *airscrew*

G. *main plane*

—————————————————————— A841

D. Flight recorders also must be capable of surviving a fuel fire of 2,000°F for 30 minutes and a smoldering fire of 500°F for 10 hours.

—————————————————————— A842

Because of their directional instability on the ground, taildraggers typically are provided with relatively large rudders. These make it possible to enter and maintain larger slip angles.

An underbelly view of a taildragger.   (A842)

$$A843$$

A blocked pitot tube. When the pitot is clogged, air pressure in the bourdon tube (diaphragm inside the instrument) remains constant. Decreasing static pressure during a climb allows the diaphragm to expand and the airspeed indication to increase. The opposite occurs during descent.

$$A844$$

B. and E. only. Air density does not change. In the case of supersonic air, airspeed *decreases*, density *increases*, and pressure *increases*. Engine inlets of a supersonic aircraft are designed to slow down incoming air to subsonic speed so that shock waves cannot enter and damage these engines.

$$A845$$

B. The others are types of flaps. The Friese aileron was designed to reduce or eliminate adverse yaw effect. When the Friese aileron moves up, a lip on the leading edge of the aileron extends beneath the bottom of the wing and increases drag on that wing.

$$A846$$

The entire flag is visible when below 10,000 feet. As the climb continues above this altitude, the flag begins to retract until it completely disappears from view at 15,000 feet. The flag is used in place of a third, or 10,000-foot, hand (needle).

$$A847$$

From geometry, a *tetrahedron* is any solid bounded by four polygons. (*Tetra* is a prefix meaning *four*, and a *polygon* is a plane figure bounded by three or more straight sides.)

$$\overline{\hspace{11cm}} A_{848}$$

B. With respect to wear alone, a tire should be replaced when any of its internal fabric is visible or when any groove is worn to its base at any spot on the tire.

$$\overline{\hspace{11cm}} A_{849}$$

True. According to avian experts and flight biologists Ken Dial and Doug Warrick, all birds can turn their heads almost 360 degrees both in flight and on the ground. They can also sustain up to 14 Gs because their heads are not above their hearts during flight.

$$\overline{\hspace{11cm}} A_{850}$$

Vertigo is a condition of the inner ear that results in a loss of balance. Spatial disorientation is a false impression of aircraft motion and/or attitude caused by misleading sensations.

$$\overline{\hspace{11cm}} A_{851}$$

Strictly speaking, spoilers are deployable surfaces that spoil or kill lift, and speed brakes are surfaces that just produce drag. Because spoilers also increase drag, they often are called *speed brakes*. Speed brakes located other than on a wing cannot be regarded as spoilers.

$$\overline{\hspace{11cm}} A_{852}$$

False. It is true on Continental engines, but the number-one cylinder on a Lycoming engine is the most forward cylinder on the right (tractor engines).

$$\overline{\hspace{11cm}} A_{853}$$

The erecting mechanism of the attitude indictor exerts a small but constant gravitational bias on the spinning gyro. This causes the gyro to maintain a right-side-up attitude with respect to the Earth's center of gravity. It is why an attitude indicator gradually biases toward a wing's-level indication during a prolonged turn.

The Peregrine falcon has been clocked in a dive at up to 220 mph. Imagine diving a Cessna 172P at V_{NE} and being passed by a bird going 38 mph faster.

True. Imagine the deceleration required to land an airplane at more than 100 mph and brake to a halt in only 20 feet, an arrest requiring less than a third of a second.

CHAPTER 8
PROCEDURES & REGULATIONS

True or False? An airplane may not be flown in IFR conditions unless a pitot-static system check has been performed within the previous 24 calendar months.

Q 857

True or False? A passenger in an unpressurized airplane must use supplemental oxygen when above 15,000 feet MSL.

Q 858

True or False? A private pilot carrying passengers in a single-engine airplane is allowed to fly 50 feet above the ocean and without flotation equipment when beyond gliding distance of land.

Q 859

True or False? An air traffic control clearance is required to enter Class D airspace.

Q 860

What is the speed limit in the Class A airspace overlying Washington, D.C.?

Q 861

What is the approximate upper limit of Class A airspace?

Q 862

Which of the following light signals does not belong?
  A. flashing white light in the air
  B. flashing white light on the ground
  C. alternating red-and-green light in the air
  D. alternating red-and-green light on the ground

A pilot is holding a magnetic heading of 185 degrees while on a VFR cross-country flight. To be legal, his cruising altitude must be

    A. 2,500 feet AGL.
    B. 3,500 feet AGL.
    C. 4,500 feet MSL.
    D. 5,500 feet MSL.

Which of the following does not belong?

    A. Charles
    B. Michael
    C. Roger
    D. Victor

The maximum-allowable compass deviation for a general aviation airplane on any given heading is

    A. 10 degrees.
    B. 20 degrees.
    C. 30 degrees.
    D. None of the above.

Which of the following does not belong?

    A. 3.8 Gs
    B. 4.4 Gs
    C. 5.7 Gs
    D. 6.0 Gs

A pilot notices solid-yellow circles that are 10 feet in diameter painted across a runway. These indicate

    A. the touchdown zone of a GPS-only approach.
    B. that emergency arresting gear is located there.
    C. that the runway is restricted to STOL operations only.
    D. that the runway is planned for permanent closure.

True or False? A pilot is approaching a long runway equipped with VASI (visual approach slope indicator) at a tower-controlled airport. He may descend at will below the glide slope indicated by the VASI.

Mix 'n' Match each of the following types of airspace with its most appropriate clue.

1. Class E airspace	A. An FSS provides service
2. Class A airspace	B. Usually extends to 4,000 feet AGL
3. Class D airspace	C. NOTAM
4. Class B airspace	D. Victor airway
5. Class C airspace	E. Ronald Reagan
6. Prohibited area	F. Not below 2,000 feet, please
7. Restricted area	G. IRs and VRs
8. Military Operations Area	H. One-mile minimum visibility at low altitude
9. Air Defense Identification Zone	I. Instrument rating required
10. Military Training Routes	J. Substantial soaring activity, for example
11. Airport Advisory Area	K. A piece of (wedding) cake
12. Alert Area	L. 200-knot speed limit
13. Class G airspace	M. Beyond the 30-mile limit
14. Special Conservation Area	N. Invisible hazards to navigation (usually)
15. Temporary Flight Restriction	O. A flight plan is required
16. Warning Area	P. Aerobatics and high-speed maneuvering

True or False? A pilot is allowed to take the entire flight test for an airline transport pilot certificate in a Cessna 152.

The empty weight and CG of an airplane must be recalculated whenever items are installed or removed. Specifically how much weight or CG change must occur before an amended weight-and-balance statement is required?

True or False? A glider pilot may fly IFR in an unpowered sailplane.

Mix 'n' Match the following:
1. Sunset to sunrise
2. End of evening twilight to beginning of morning twilight
3. One hour after sunset to one hour before sunrise

A. May log night time
B. Position lights required
C. Three takeoffs and landings needed for currency

Which of the following is not required for a night IFR flight in a conventionally equipped Piper Warrior?
A. clock
B. gyroscopic rate-of-turn indicator
C. vertical-speed indicator
D. slip-skid indicator

Mix 'n' Match the following parts of the Federal Aviation Regulations with their most appropriate clues:

1. Part 1	A. myocardial infarction
2. Part 23	B. Trans World Airlines
3. Part 25	C. decibels
4. Part 36	D. Cessna 182RG
5. Part 61	E. less than 254 pounds
6. Part 67	F. Embry Riddle Aero. Univ.
7. Part 91	G. Noah Webster
8. Part 103	H. three landings every 90 days
9. Part 121	I. charter flights
10. Part 135	J. Boeing 777
11. Part 141	K. 1,000 feet and 3 miles

Which of the following does not belong?
   A. 121.5 MHz
   B. 243.0 MHz
   C. 406.0 MHz
   D. 500.0 MHz

If a pilot were to inadvertently squawk 5200 on his transponder, he could be mistaken by an air traffic controller for a
   A. remotely piloted vehicle.
   B. pilot who has declared "minimum fuel."
   C. truck carrying hazardous material.
   D. ship at sea.

True or False? The VHF frequency band ranges from 20 to 200 MHz.

Class A airspace begins at Flight Level 180 (approximately 18,000 feet MSL). It requires that a pilot have an instrument rating and that his airplane be IFR equipped. How is it possible for a VFR-only pilot to fly at FL240 over the United States without an IFR-equipped airplane, without a clearance, and without special permission?

Q 880

True or False? A pilot may drop a bowling ball from his aircraft.

Q 881

The letter "N" is used to designate U.S.-registered aircraft. Only eight other countries use a single letter for this purpose. Name the eight countries represented by these letters:

1. B _____________		5. G _____________
2. C _____________		6. I _____________
3. D _____________		7. P _____________
4. F _____________		8. Z _____________

Q 882

Who does not belong?

    A. Tom Cruise
    B. Arnold Schwarzenegger
    C. Patrick Swayze
    D. John Travolta

Q 883

A VFR pilot with seaplane and landplane ratings takes off from a lake in an amphibian, flies for an hour, and lands on a concrete runway. How must the flying time be logged?

    A. all as seaplane time
    B. all as landplane time
    C. half seaplane, half landplane
    D. It does not matter.

____ **Q 884**

True or False? A VFR pilot is 99 NM from his destination and is passing over an airport. A large bug enters his pitot tube and renders his airspeed indicator inoperative. The pilot may continue to his destination.

____ **Q 885**

True or False? It is legal for a pilot to swing an aircraft compass.

____ **Q 886**

True or False? An American pilot with a U.S. private pilot certificate and fewer than 200 hours of flying time wants to fly an Aerospatiale Trinidad that is in the United States but registered in France. He may not fly this airplane as pilot-in-command.

____ **Q 887**

A pilot is taxiing for takeoff on a snow-covered taxiway. He knows that he has reached the correct runway because of the red-and-white sign bearing the runway number, but the hold-short lines are obscured. He must hold

    A. 25 feet from the edge of the runway.
    B. 50 feet from the edge of the runway.
    C. so that no part of the airplane passes beyond the edge of the runway.
    D. so that no part of the airplane passes beyond the sign.

____ **Q 888**

True or False? It is legal for a pilot to land in a crosswind that exceeds the maximum demonstrated crosswind as published in the pilot's operating handbook.

____ **Q 889**

True or False? It is legal for a VFR-only pilot to fly extended distances in a small airplane above a solid undercast with no more than the instruments required for ordinary VFR flight.

True or False? A private pilot takes his 10-year-old daughter for a ride in the family airplane and allows the girl to fly the airplane during cruise flight. The private pilot may log the cruise portion of this flight as pilot-in-command.

A departing pilot requests and obtains approval from the tower controller to make a straight-out departure. The pilot should not turn until

   A. outside Class D airspace.
   B. beyond the traffic pattern.
   C. 3 statute miles beyond the end of the runway.
   D. 3 nautical miles beyond the end of the runway.

An instrument-rated pilot is flying IFR in Class G airspace and is not communicating with air traffic control. He should squawk which of the following codes on his transponder?

   A. 1200
   B. 3100
   C. 5100
   D. a discrete code

True or False? A pilot may stress a typical lightplane in the Normal category to as much as +3.8 Gs (the normal, positive, limit load factor) except when the wing flaps are extended.

True or False? According to the *Aeronautical Information Manual*, the tarmac is defined as all paved aircraft movement areas excluding taxiways and runways.

True or False? A flashlight (but not spare batteries) is required equipment aboard an aircraft being flown at night by a private pilot.

A pilot is proceeding toward an airport that is surrounded on a sectional chart by a magenta, dashed, keyhole-shaped line. The airport is situated within or under

A. Class D airspace beginning at the surface.
B. Class D airspace beginning at 700 feet AGL.
C. Class E airspace beginning at 700 feet AGL.
D. Class E airspace beginning at the surface.

True or False? A pilot approaches a small house while flying over an otherwise flat, featureless desert. It is legal for him to fly 400 feet abeam the house at an altitude of 400 feet AGL.

What is missing from the following list of VFR maneuvers?

A. touch-and-go landing
B. low approach
C. missed approach
D. full-stop landing

Which of the following does not belong?

A. a pint of 150-proof liquor
B. a cigarette lighter with a liquid reservoir
C. an animal with an implanted heart pacemaker
D. 2.5 kilograms of dry ice

When exiting the runway after landing at a controlled airport, the airplane is not considered clear of the runway until

A. all of the aircraft has passed beyond the runway shoulder line.

B. all of the aircraft has passed beyond the edge of the runway.

C. part of the aircraft has crossed the double yellow lines.

D. all of the aircraft has passed beyond the double yellow lines.

The battery of an emergency locator transmitter (ELT) must be replaced after it has been used for more than _______ minutes or after it has been installed for more than _______ percent of its useful life.

The two frequencies designated for air-to-air communications between aircraft are _______ and _______ MHz.

A pilot is at an indicated altitude of 8,000 feet with his encoding altimeter set to 30.12 at a time when it should be set to 30.32. An air traffic controller viewing the transponder return on his radar screen will believe the airplane to be at

A. 7,800 feet.
B. 8,000 feet.
C. 8,200 feet.
D. 8,400 feet.

When an aircraft accident report is required, it must be made to the FAA within

A. 24 hours.   C. 72 hours.
B. 48 hours.   D. Never.

_____________________ **Q 905**

True or False? If a flight instructor loses his medical certificate, there are times when he may still provide required flight instruction and charge for his services.

_____________________ **Q 906**

The pilot of Cessna 206 Stationair N222LS is transporting urgently needed lifesaving material (such as a human heart). What call sign should she use to indicate the need for expeditious handling by air traffic control?

_____________________ **Q 907**

True or False? There is no minimum age for a pilot wanting to fly from Seattle to Miami in a powered aircraft with an empty weight of 250 pounds.

_____________________ **Q 908**

True or False? While preflighting his engine, a pilot notices that the electric fuel pump is labeled, "For Automotive Use Only." The flight must be postponed until the pump is replaced.

_____________________ **Q 909**

True or False? The top of an airport's Class D airspace is shown on the chart as "-27." This means that a pilot may fly above the airport at 2,700 feet MSL without contacting the control tower.

_____________________ **Q 910**

True or False? A private pilot with multi-engine and instrument ratings but no jet training may serve as second-in-command (co-pilot) of a small business jet that requires two pilots if the aircraft is not being operated for hire.

When using a VASI (visual approach slope indicator), a pilot is assured safe obstruction clearance only when he is within _______ degrees of the extended runway centerline and is no more than _______ NM from the runway threshold.

What is the minimum safe altitude (AGL) when overflying an area designated on a sectional chart by a parachute symbol?

True or False? Aerobatic maneuvers may not be conducted in aircraft that are not certificated in the aerobatic category.

Fill in the blanks: When _______________, Class ___ airspace reverts to Class ___. And when _______________, it reverts further to Class ___ airspace.

According to 14 CFR Part 91, all aircraft are required to display position lights when flown at night. How are position lights arranged on a balloon?

A pilot flying an airplane on a magnetic heading of 360 degrees sees a balloon that is tracking 270 degrees ahead and to his right. Both aircraft are at the same altitude. According to regulation, the pilot of the airplane must

   A. climb.
   B. descend.
   C. alter course to the right.
   D. None of the above.

____ **Q 917**

A VFR pilot approaching an airport is beneath a flat, level overcast while observing cloud-clearance requirements, and there are no clouds at or below his altitude. As he descends to enter the traffic pattern, however, he violates cloud-clearance requirements. How is this possible?

____ **Q 918**

A pilot is about to embark on a high-altitude flight and plans to use a portable oxygen system that includes a steel oxygen tank. For the system to be used legally, it must have been pressure-tested within the previous

    A. 3 years.
    B. 5 years.
    C. 8 years.
    D. 10 years.

____ **Q 919**

On different occasions, a pilot keys his microphone 3, 5, and 7 times to activate pilot-controlled airport lighting. The runway-edge lights will remain illuminated for ________, ________, and ________ minutes, respectively.

____ **Q 920**

An airworthiness certificate is normally white. What is the significance of one that is pink?

____ **Q 921**

While landing at an airport with standard runway-edge lights, a pilot touches down abeam a runway light to his left and comes to a stop nine runway lights thereafter. The length of the landing roll is approximately

    A. 900 feet.
    B. 1,350 feet.
    C. 1,800 feet.
    D. Cannot be determined.

A fire truck pulls alongside and in front of a taxiing airplane that had just made an emergency landing. A firefighter steps off the truck and makes a beckoning movement with his left hand and arm. What should the pilot do?

The most current general aviation transceivers have ________ channels that cover the VHF frequency band from ________ to ________ MHz.

May a tower controller allow an airplane to land on a runway that is occupied by a preceding airplane that has just landed?

Arrange the following alphabetically listed aircraft in the order in which they have right-of-way according to Part 91 of the Federal Aviation Regulations:

    A. airplane
    B. airplane towing a glider
    C. balloon
    D. blimp
    E. glider
    F. helicopter

Which of the following does not belong?

    A. throwing chaff out of a window
    B. dropping a flare
    C. squawking 7600
    D. flying a triangular pattern

Q 927

True or False? It is legal for an instrument-rated private pilot to take off in a single-engine airplane at night when the ceiling and visibility are zero.

Q 928

True or False? A VFR pilot experiencing two-way communications failure may enter Class D airspace, circle the airport, and wait for a light signal from the control tower.

Q 929

An air traffic controller tells a pilot to "Contact Albuquerque Center on 121.475 MHz." How would the pilot know that he has been given an incorrect frequency?

Q 930

When may a pilot perform a loop in a properly certificated airplane without wearing a parachute?

   A. when flying solo
   B. when receiving dual instruction
   C. Both of the above.
   D. None of the above.

Q 931

A pilot flying in VFR conditions enters the traffic pattern at an airport with an elevation of 200 feet and a published traffic pattern altitude of 1,200 feet at a time when the tower controller reports a 1,300-foot overcast. The downwind leg should be flown at altitude of

   A. 1,200 feet MSL.
   B. 1,100 feet MSL.
   C. 1,000 feet MSL.
   D. 800 feet MSL.

What restrictions do the Federal Aviation Regulations place on aerobatic flight at night?

True or False? A pilot has not flown for more than 90 days. He then makes three solo takeoffs and landings at night. He may now carry passengers during the day.

Which of the following does not belong?
   A. air traffic controller
   B. dispatcher
   C. parachute rigger
   D. fueler

True or False? It is illegal to land on a runway that is closed and marked by a large "X" at each end.

True or False? It is legal to land on roads and highways in Alaska for any reason.

Which of the following does not belong?
   A. Repair broken landing light circuits.
   B. Replace, clean, and gap spark plugs.
   C. Replace fuel-line connections.
   D. Replace hydraulic-hose connections.

True or False? All airports with an operating control tower are in either Class B, C, or D airspace.

Under what conditions might a tower controller instruct a pilot to "begin a high-speed taxi?"

When an FAA-certificated pilot flies his U.S.-registered airplane outside the United States (including Canada and Mexico), he must abide by

A. the Federal Aviation Regulations.
B. regulations of the International Civil Aviation Organization (ICAO).
C. the regulations of the host country.
D. All of the above.

The only four instruments required on a hot-air balloon are an altimeter, a vertical-speed indicator, a fuel gauge, and a ________.

Why is a compass not required on a balloon?

Basic VFR minimums when operating above 10,000 feet MSL require a minimum visibility of ________; minimum distance from clouds is ________ feet above, ________ feet below, and ________ feet horizontally.

True or False? There are times when the minimum visibility for VFR flight in Class G (uncontrolled) airspace below 10,000 feet MSL is 3 SM.

True or False? Other than during an emergency, a pilot may not operate an aircraft in a careless or reckless manner.

With respect to a typical general aviation airplane, which of the following does not belong?

A. damage to property, other than the aircraft, in excess of $25,000

B. failure of the structural components of an engine

C. in-flight fire

D. flight control system malfunction or failure

A pilot lands on a runway with white runway centerline lighting during poor visibility. When the white lights begin to alternate with red, the pilot knows that there are _______ feet of runway remaining. When the alternating white and red lights become red only, there are only _______ feet of runway remaining.

A. 3,000; 1,500

B. 3,000; 1,000

C. 2,500; 1,500

D. 2,000; 1,000

True or False? A private pilot is about to take off in a typical single that is not equipped with a minimum-equipment list. He discovers that one of his two VOR receivers is inoperative. He may legally depart on a lengthy, VFR cross-country flight at night.

True or False? The minimum legal age for soloing an aircraft in the United States is 14.

True or False? A pilot is concerned about the safety of his flight and needs timely but not immediate assistance. This is an emergency and justifies emergency action.

True or False? According to Part 91 of the Federal Aviation Regulations, an aerobatic maneuver is one in which pitch angle exceeds 30 degrees and bank angle exceeds 60 degrees.

Which of the following does not belong?
- A. black seal
- B. blue seal
- C. gold seal
- D. red seal

After takeoff, a pilot decides to remain in the traffic pattern. He should not begin his turn onto the crosswind leg until
- A. within 300 feet of pattern altitude.
- B. within 500 feet of pattern altitude.
- C. beyond the departure end of the runway.
- D. one-quarter mile beyond the end of the runway.

After takeoff, a pilot decides to exit the traffic pattern. He should not make his first turn until
- A. within 300 feet of pattern altitude.
- B. reaching pattern altitude.
- C. beyond the departure end of the runway.
- D. one-quarter mile beyond the end of the runway.

True or False? The pilot-in-command of any certified, civil, U.S.-registered aircraft must use seat belts during all takeoffs and landings.

____ **Q 956**

An instrument pilot on an IFR flight plan lifts off from Runway 24, which has a magnetic direction of 245 degrees, under the influence of a crosswind that causes a 5-degree left drift angle. He is advised by air traffic control to "maintain runway heading after takeoff." His initial magnetic heading should be ______ degrees.

____ **Q 957**

A discrete transponder code is issued to a pilot so that air traffic control can easily identify his aircraft on secondary radar. All 4,096 codes may be used as discrete codes except ________.

____ **Q 958**

True or False? The Federal Aviation Regulations permit an instrument-rated pilot flying a properly certificated sailplane (not a motorglider) to request a clearance from air traffic control and execute an ILS approach during actual instrument conditions.

____ **Q 959**

A pilot lands on a long runway at night. During a normal landing roll, he crosses a row of pulsing white lights imbedded in the runway. What is the significance of these lights?

____ **Q 960**

True or False? A pilot who is legally or totally blind in one eye may be issued a first-class, second-class, or third-class medical certificate.

____ **Q 961**

For instrument-rated pilots: Technically speaking, a pilot is required to time that portion of an ILS approach between the final-approach fix and

    A. decision height.
    B. the missed-approach point.
    C. the runway threshold.
    D. Timing is not required.

Which of the following does not belong?

  A. NOTAM (D)
  B. NOTAM (L)
  C. NOTAM (M)
  D. FDC NOTAM

Which of the following does not belong?

  A. 122.7 MHz
  B. 122.8 MHz
  C. 122.9 MHz
  D. 123.0 MHz

True or False? A pilot on a VFR cross-country flight during the hours of darkness arrives at his destination during daylight hours (after sunrise). He is required to land with a minimum fuel reserve of only 30 minutes.

For instrument-rated pilots: The maximum-allowable airspeed while holding in a piston-powered, general aviation airplane at 9,000 feet MSL is

  A. 175 KIAS.
  B. 200 KIAS.
  C. 230 KIAS.
  D. 265 KIAS.

_______________________________________ **Q 966**

A pilot is flying VFR above the 5,000-foot-high floor of a
Colorado valley while on a magnetic heading of 179 degrees and
under the influence of an easterly wind. Which of the following
cruise altitudes may not be used?

    A. 6,500 feet
    B. 7,500 feet
    C. 8,500 feet
    D. 9,500 feet

_______________________________________ **Q 967**

A pilot is flying VFR at night in Class G (uncontrolled) airspace
at 10,500 feet MSL and more than 1,200 feet AGL. The minimum
visibility required for this flight is _________ statute miles, and the
cloud clearance requirements are the same as when flying VFR
during the day and at the same altitude as in Class _________ airspace.

    A. 1, B   D. 3, E
    B. 1, E   E. 5, B
    C. 3, B   F. 5, E

_______________________________________ **Q 968**

True or False? With an elevation of 9,927 feet MSL, Leadville,
Colorado is the highest airport in the United States with an
FAA-approved instrument approach.

_______________________________________ **Q 969**

True or False? The shortest (airplane) runway for which an
instrument approach may be approved is 2,200 feet.

_______________________________________ **Q 970**

True or False? When installed in an aircraft used for IFR flight,
the attitude indicator and turn coordinator (or turn indicator)
must be powered by independent sources. In other words, one
must be vacuum (or pressure) powered while the other must be
electrically powered.

____ **Q 971**

A pilot overflying desolate terrain notices a downed airplane with two people waving near the wreckage. A large letter X has been carefully trampled by the survivors in the snow near the wreckage. This signals that

    A. assistance is already on the way.
    B. a landing should not be attempted there.
    C. food and water are required.
    D. medical assistance is required.

____ **Q 972**

True or False? Manifold pressure gauges are required on airplanes powered by piston engines and equipped with constant-speed propellers.

____ **Q 973**

A pilot departs Reno, Nevada on a VFR flight plan to Santa Monica, California at 1940 UTC. She makes an en route position report at 2030 during which she estimates her next reporting fix at 2110. She also changes the ETA for her destination from 2240 to 2225 UTC. The pilot, however, is not heard from again. In this case, a telephonic search for the aircraft would begin at _______ UTC. If it were not located in this manner, a decision to initiate search-and-rescue activities would be made at _______ UTC.

____ **Q 974**

An emergency locator transmitter (ELT) should always be capable of sustaining a continuous broadcast for at least _______ hours.

____ **Q 975**

True or False? When clearing a runway after landing at a controlled airport, a pilot may switch to ground control without waiting to be advised by the tower controller to do so.

Pilots are advised to abide by the Five C's when lost. What are the Five C's, and can you name them in their proper sequence?

What are the two most common reasons for a runway to have a displaced threshold?

According to the *Aeronautical Information Manual*, a pilot should complete the turn from base leg to final approach at least ______________________ from the runway threshold.

   A. one-eighth mile

   B. one-quarter mile

   C. one-half mile

   D. The AIM does not specify a minimum length for final approach.

True or False? A pilot is operating at 2,500 feet AGL at a distance of 3.5 NM from a controlled airport that has an elevation of 2,800 feet. The Class D airspace surrounding the airport has a radius of 3.0 NM and an upper limit of 5,000 feet MSL. The speed limit for that aircraft is 250 knots.

True or False? A private pilot not current at night is legally qualified to land a passenger-carrying airplane at 1920 (local time) on a day when sunset occurs at 1830 and the end of civil twilight occurs at 1900.

According to the *Aeronautical Information Manual,* pilot impairment contributes to many more accidents than do aircraft system failures. It recommends using the acronym, "I'M SAFE," as a checklist of possible personal impairments that could compromise safety. What potential problem does each letter represent?

A pilot exercises his emergency authority and intentionally violates one or more of the Federal Aviation Regulations in Part 91, *General Operating and Flight Rules.* He is required to file a report of this (these) violation(s) to the FAA

    A. within 24 hours.
    B. within 7 days.
    C. within 30 days.
    D. only upon request.

True or False? The non-stop, around-the-world flight of the *Voyager* could not be logged as a cross-country flight.

A pilot overhears an air-carrier pilot transmitting, "Los Angeles Tower, this is United Thirty Six, Lufthansa, Boeing 747, over." What is the meaning of this seemingly contradictory transmission?

A piston-powered airplane and a turbine-powered airplane are being operated under Part 91 of the Federal Aviation Regulations. According to regulation, the piston engine ________ be overhauled when it reaches TBO (time between overhauls), and the turbine engine ________ be overhauled when it reaches TBO.

    A. does not have to, must
    B. does not have to, does not have to
    C. must, must
    D. must, does not have to

The FAA requires that VASI equipment be capable of generating visual glideslopes as shallow as 1 degree and as steep as

A. 6 degrees.
B. 9 degrees.
C. 12 degrees.
D. 15 degrees.

While listening to an air traffic control frequency, you overhear a pilot identifying himself as "Tango-Bonanza-Four-Six-Four-Alpha." The use of *tango* indicates that the aircraft

A. has a turbine engine.
B. has two engines.
C. is an air-taxi flight.
D. is turbocharged.

True or False? All tower-controlled civil airports that provide weather service are in Class B, C, or D airspace (when the control tower is open).

A pilot about to embark on a winter flight determines that the barometric pressure en route is forecast to be in excess of 31.00 inches Hg. What must he do prior to operating in such a high-pressure area?

What is the call sign given to a general aviation aircraft in which the current U.S. president is flying?

A pilot transmitting on an ATC frequency identifies his aircraft as "Compassion Three-Seven-Golf." What does such a call sign signify?

True or False? All civil airports in the United States equipped with runway lights are also equipped with rotating beacons.

True or False? Other than in the vicinity of certain airports, the base of all Class E airspace in the United States is at either 700 or 1,200 feet AGL.

What is the booze news?

True or False? The minimum safe altitude for an airplane over a congested area is 1,000 feet above the highest obstacle within a horizontal radius of 2,000 feet.

True or False? An FAA-certificated pilot may log flight time as pilot-in-command of an ultralight airplane.

Several lightplanes are not permitted to fly above their operational ceilings, which are below their service and absolute ceilings. Why are they so limited?

A pilot flying a typical single with a fixed-pitch propeller enters a dive and accelerates to V_{NE} with the *throttle fully closed*. He should anticipate that engine rpm will exceed the redline.

Polish Ogar motorglider.   (Q999)

_______________________________________ **Q 999**

True or False? A certain private pilot has only one rating: airplane single-engine land. This pilot, therefore, may not act as pilot-in-command of a motorglider but may carry passengers in an amphibian.

_______________________________________ **Q 1,000**

Every nation has sovereign control of its airspace, but to what altitude does this sovereignty extend?

_______________________________________ **Q 1,001**

True or False? Prior to being certified, a new lightplane must be flown substantially faster than V_{NE}, the never-exceed (redline) airspeed.

False. A pitot-system check is not required.

False. Supplemental oxygen must be provided to passengers when above 15,000 feet, but they are not required to use it.

True. This proves again that what is legal is not necessarily safe.

False. Permission, not a clearance, is required to enter Class D airspace. A clearance is required, however, to enter Class A and B airspace.

The speed limit in Class A airspace over Washington, DC is the same as it is in Class A airspace everywhere else over the United States. It is Mach 1.0, the speed of sound.

B. The upper limit of Class A airspace is Flight Level 600, which is approximately 60,000 feet MSL. Class E airspace begins above that and continues to some unspecified altitude.

A. A flashing white light from the tower has no meaning to a pilot in flight. On the ground, it indicates that the pilot should return to his starting point on the airport.

$$A863$$

A. Cruise altitude is determined by magnetic course. Since this aircraft has a magnetic heading of 185 degrees, and the amount of drift (or wind correction) is unknown, there is no way to know if the magnetic course being flown requires either an even +500 or an odd +500 cruise altitude. Consequently, the only altitude that is guaranteed legal is one that is below 3,000 feet AGL where there is no required relationship between altitude and direction of flight.

$$A864$$

C. Three of these gentlemen have their names (Charlie, Mike, and Victor) included in the International Phonetic Alphabet. Roger is not so honored.

$$A865$$

A. The use of a particular piece of equipment may be allowed to cause compass deviation in excess of 10 degrees, but a warning placard must be conspicuously displayed at such a time.

$$A866$$

C. The minimum limit load factors for normal, utility, and aerobatic aircraft certificated under Part 23 of the Federal Aviation Regulations are 3.8 Gs, 4.4 Gs, and 6.0 Gs, respectively.

$$A867$$

B. This would be an interesting way to prevent an overrun but might be hard on the propeller.

$$A868$$

False. A pilot may not descend below the glide slope—as defined by the VASI—at a towered airport unless a lower altitude is necessary for a safe landing. Refer to 14 CFR §91.129(e)(3).

1. D	9. O
2. I	10. G
3. L	11. A
4. K	12. J
5. B	13. H
6. E	14. F
7. N	15. C
8. P	16. M

A 870

True. The aircraft is required to have the necessary avionics. The successful applicant is issued an ATP certificate with a single-engine, land rating.

A 871

The weight change must equal or exceed one-half of 1 percent of the maximum-allowable landing weight, or the CG must change by at least one-half of 1 percent of the mean aerodynamic chord.

A 872

True. To fly IFR, however, the glider must be properly equipped, and the pilot must have both a glider rating and an instrument rating valid in airplanes.

A 873

1. B
2. A
3. C

A 874

C. A VSI is not required for IFR flight. (A turn indicator is not required if the airplane is equipped with three attitude indicators.)

1. G. *Definitions and Abbreviations*
2. D. *Airworthiness (propeller airplanes)*
3. J. *Airworthiness (transport category)*
4. C. *Noise Standards*
5. H. *Certification: Pilots and Instructors*
6. A. *Medical Standards and Certification*
7. K. *General Operating and Flight Rules*
8. E. *Ultralight Vehicles*
9. B. *Certification and Operations (major airlines)*
10. I. *Air Taxi and Commercial Operators*
11. F. *Pilot Schools*

D. The first two frequencies are transmitted by a conventional ELT. The third is transmitted by the new-generation ELT.

D. Large ships at sea often are assigned code 5200. Transponders are not just for airplanes.

False. The VHF frequency band ranges from 30 to 300 MHz.

Because there is no Class A airspace over the Hawaiian Islands, a VFR pilot may fly as high there as his airplane will take him.

True. According to Federal Aviation Regulation 91.15, a pilot may drop any object as long as "reasonable precautions are taken to avoid injury or damage to people or property."

____ A 881

B = China, C = Canada, D = Germany, F = France, G = United
Kingdom, I = Italy, P = North Korea, and Z = Zimbabwe.

____ A 882

B. Of the four, only Schwarzenegger is not a licensed pilot.

____ A 883

D. Once a pilot is rated in a class of aircraft, there is no requirement to log flight time in that particular class. He might, however, need to log a seaplane takeoff and a landplane landing to
   comply with the recent flight experience requirements.

Lake LA-4 amphibian about to enter the water.   (A883)

$$A884$$

False. According to 14 CFR §91.7(b), a pilot "shall discontinue a flight when unairworthy mechanical, electrical, or structural conditions occur."

$$A885$$

True. A pilot, however, may not compensate (adjust the magnets) of a compass.

$$A886$$

False. He may fly this airplane as pilot-in-command. Refer to 14 CFR §61.3(b).

$$A887$$

D. This is according to the FAA's uniform signage program.

$$A888$$

True. The maximum demonstrated crosswind is simply the maximum crosswind component that was available to test pilots during the certification process of the airplane.

$$A889$$

True. This does not mean that such an operation is necessarily safe.

$$A890$$

False. Refer to 14 CFR §61.51(e).

$$A891$$

There is no legal definition for a straight-out departure. A pilot should use his best judgement.

A892

A. Air traffic controllers observing your transponder code will not know or care that the pilot is operating in IFR conditions.

A893

True. With the flaps fully extended, the limit load factor in most cases is limited to +2.0 Gs.

A894

False. Tarmac (a registered trademark) is similar to tarmacadam (also known as blacktop), a type of pavement made by pouring tar over crushed stone.

A895

False. Not having a flashlight available at night, however, would be foolish.

A896

D. Otherwise, Class E airspace normally begins at either 700 or 1,200 feet AGL.

A897

True. According to 14 CFR §91.119(c), an aircraft flying over sparsely populated areas "may not be operated closer than 500 feet to any…structure." In this case, the aircraft is 565 feet (diagonally) from the house.

A898

Stop-and-go landing. When a pilot is "cleared for the option" by a tower controller, he may exercise any of these five options (according to the *Aeronautical Information Manual*).

_______________________________________ A899

C. According to 14 CFR Part 175, which deals with the carriage
   of hazardous materials, the other items may not be taken
   aboard commercial flights (including a charter flight in a
   Cessna 152).

_______________________________________ A900

D. These lines also are called the hold-short lines (or the holding
   position markings).

_______________________________________ A901

60 minutes, 50 percent

_______________________________________ A902

122.75 and 122.85 MHz

_______________________________________ A903

C. The actual altitude of the airplane is 8,200, which is the
   altitude shown on the controller's radar screen (irrespective
   of the altimeter setting used by the pilot).

_______________________________________ A904

D. Regulations require that accident reports be made to the
   nearest National Transportation Safety Board field office (and
   not the FAA) as expeditiously as possible.

_______________________________________ A905

True. A CFI is allowed to instruct and charge a fee without a
medical certificate as long as the "student" is qualified to act as
pilot-in-command of the aircraft being used.

Answers

—————————————————————— A906

Lifeguard Two Two Two Lima Sierra.

—————————————————————— A907

True. This is according to 14 CFR Part 103, which deals with ultralight vehicles.

—————————————————————— A908

Not necessarily. Beech and Cessna have used such fuel pumps (with FAA approval) in many of their aircraft.

—————————————————————— A909

True. The "-27" indicates that the Class D airspace extends up to but does not include 2,700 feet.

—————————————————————— A910

False. The pilot must meet the currency requirements of 14 CFR §61.55(b).

—————————————————————— A911

10 degrees and 4 nautical miles.

—————————————————————— A912

There is none (except when operating in Class A, B, and C airspace and being advised by air traffic control about skydiving operations). Otherwise, it is suggested that pilots give parachute jumping areas a wide berth.

—————————————————————— A913

False. Limited aerobatic maneuvers often are permitted when operating some Normal-category airplanes in the utility category when the pilot complies with the limitations published in the pilot's operating handbook.

____ A914

the tower is closed, D, E, weather reporting is unavailable, G

____ A915

A balloon is required only to display a single white position
light that must be visible from all directions. This is in addition
to a single flashing red or white anticollision light.

____ A916

D. The pilot may continue straight ahead because the two aircraft
are in no danger of colliding. Think about it.

____ A917

The pilot begins descending from the floor of Class B airspace
where he has been 100 feet beneath the overcast. This is legal
because he is required only to remain clear of clouds in Class B
airspace. He descends into the top of underlying Class D airspace
where he suddenly is required to be 500 feet beneath that same
overcast.

____ A918

B. Aluminum and composite cylinders must be tested every 3
years (according to regulations of the Department of Transportation). In all cases, the oxygen bottles are stamped with
the date of the last test. Similar pressure tests are required of
installed systems.

____ A 919

15, 15, and 15. Varying the number of microphone
clicks has no effect on how long the lights remain on. It only
adjusts lighting intensity when such a feature is available.

373

Answers

$$A 920$$

A pink airworthiness certificate is issued to aircraft that are either experimental or restricted. It is not to be confused with a pink registration certificate, which is an interim or temporary certificate to be used until a permanent one is issued.

$$A 921$$

C. Landing lights are typically and uniformly spaced at intervals of 200 feet, which makes them useful for estimating (poor) visibility and approximating landing and takeoff distances.

$$A 922$$

This signal means that aircraft evacuation is recommended, probably because the firefighter detects a hazardous condition. This and other important fire fighting and rescue instructions for pilots are shown in the *Aeronautical Information Manual*.

$$A 923$$

760, 118.000, 136.975

$$A 924$$

Yes, but only during daylight hours and when a safe distance separates the airplanes.

$$A 925$$

Balloon, glider, airplane towing glider, blimp, and helicopter or airplane. (Helicopters and airplanes have equal rights-of-way with respect to each other.)

$$A 926$$

B. The other actions either were or are procedures to be used by a pilot in instrument conditions to notify a radar controller of two-way communications failure.

_______________________________________ **A**927

True. This is more proof that what is legal is not necessarily safe.

_______________________________________ **A**928

True. According to the *Aeronautical Information Manual*, the pilot should remain outside Class D airspace until determining traffic direction and flow. He may then join the traffic pattern and watch for a light signal.

_______________________________________ **A**929

The emergency communications frequency is protected by a buffer that excludes using three channels immediately above and below 121.5 MHz.

_______________________________________ **A**930

A. An instructor and student are not required to wear parachutes when practicing an aerobatic maneuver (such as a spin) required for an FAA flight test. Because a loop is not so required, parachutes must be worn.

_______________________________________ **A**931

C. The base of the overcast is 1,500 feet MSL. Being in controlled airspace (Class D), the pilot must observe cloud-clearance regulations.

_______________________________________ **A**932

None. The same rules apply to aerobatics performed at night as when performed during the day.

_______________________________________ **A**933

True. The regulations require only that the pilot have made three takeoffs and landings to legally carry passengers during the day. Daytime or nighttime conditions are not specified.

—————————————————————————— A934

D. The other three require FAA certification.

—————————————————————————— A935

False. A pilot landing on a closed runway might, however, be in violation of 14 CFR §91.13 (careless or reckless operation).

—————————————————————————— A936

True. According to the Alaska Airmen's Association, the only restriction is that the pilot not disrupt traffic.

—————————————————————————— A937

D. 14 CFR Part 43 defines the other tasks as being allowed under the provisions of preventative maintenance as long as they do not involve complex assembly operations.

—————————————————————————— A938

False. According to the *Aeronautical Information Manual*, if the control tower does not provide weather reporting (a requirement for surface-based controlled airspace), it is in Class G airspace (although it might underlie Class E airspace).

—————————————————————————— A939

The instruction might be issued to an aircraft holding in position on the runway and immediately prior to a takeoff clearance to make room between an aircraft on short final and one that is slow to clear the runway.

—————————————————————————— A940

D. When any of these regulations are in conflict, the pilot must abide by the most restrictive ones (per 14 CFR §91.703). It is the pilot's responsibility to know the applicable regulations.

A pyrometer, which measures the temperature of the hot air at the top of the balloon.

A 941

A balloon is free to rotate about its vertical axis, and its pilot cannot maintain a given heading. A compass is useful, however, in approximating track when movement over the ground can be detected.

A 942

5 statute miles, 1,000 feet, 1,000 feet, and 5,280 feet (one statute mile).

A 943

True. The minimum visibility in Class G airspace is 1 mile during daylight hours and 3 miles at night.

A 944

False. Although extremely discouraged, a pilot may fly in a careless or reckless manner as long as he does not endanger the life or property of another.

A 945

B. The other occurrences must be immediately reported to the National Transportation Safety Board by the pilot or operator.

A 946

B. Such a centerline lighting system is usually associated only with runways served by an ILS (instrument landing system).

A 947

$$A 948$$

True. However, the receiver must not be required by the aircraft's equipment list, an airworthiness directive, or a supplemental type certificate. Also, it must be removed before departure or deactivated and placarded "inoperative."

$$A 949$$

True. This applies, however, only to balloons and gliders. A student must be 16 years old to solo an airplane or helicopter. There is no minimum age for soloing an ultralight.

$$A 950$$

True. Such an urgent condition is defined in the *Aeronautical Information Manual* as an emergency.

$$A 951$$

False. 14 CFR §91.303 defines an aerobatic maneuver as one involving an abrupt change in attitude, an abnormal attitude, or abnormal acceleration.

$$A 952$$

D. A blue seal on an airman certificate indicates demonstrated instrument proficiency; a gold seal on a flight instructor certificate indicates a specified level of high achievement; a black seal emblazons all other airmen certificates.

$$A 953$$

A. and C. So sayeth the *Aeronautical Information Manual*.

$$A 954$$

B. and C. So sayeth the *Aeronautical Information Manual*.

$$A\,955$$

False. Pilots of most balloons and some airships are not required to abide by this regulation.

$$A\,956$$

245 degrees. The pilot is expected to maintain a magnetic heading that coincides with the magnetic direction of the runway.

$$A\,957$$

Any code ending in double-zero (such as 1200, 5600, 7700, etc.).

$$A\,958$$

True. According to FAA's Flight Standards Division, he may execute the approach if he is assured of not having to hold or maintain an assigned altitude, is certain that he will not have to execute a missed approach (that would be interesting), and is sufficiently glib to avoid being charged with a careless or reckless operation.

$$A\,959$$

These land-and-hold-short lights are installed across certain runways approved for land-and-hold-short operations (LAHSO) at the hold-short point (to help pilots avoid tangling with aircraft operating on an intersecting runway).

$$A\,960$$

True. As of January, 1997, there were 184, 638, and 2,169 monocular pilots holding first-class, second-class, and third-class medical certificates, respectively. Wiley Post is the most famous of all monocular pilots.

$$A\,961$$

D. Timing is required only during certain nonprecision approaches.

_______________________________________ A962

C. A (D)istant NOTAM relates to safety of flight, a (L)ocal
   NOTAM usually is not critical to safety, and an FDC (Flight
   Data Center) NOTAM relates to changes of a regulatory
   nature. There is no such thing as a NOTAM (M).

_______________________________________ A963

C. The others are conventional UNICOM frequencies used at
   airports without operating control towers. 122.9 MHz is a
   MULTICOM frequency that is used for more specific purposes.

_______________________________________ A964

False. There is no minimum fuel reserve required at the *time* of
landing. 14 CFR §91.151 is a *planning* requirement, which means
that fuel reserves must be in the tank at the beginning of the flight.
Although imprudent, this reserve may be consumed en route.

_______________________________________ A965

C. Maximum-allowable holding speeds for any aircraft is 200
   KIAS (up to and including 6,000 feet), 230 KIAS (6,001
   through 14,000 feet), and 265 KIAS (above 14,000 feet).

_______________________________________ A966

D. The first two altitudes may be used because the aircraft would
   be less than 3,000 feet AGL, and the hemispherical altitude
   rule would not be applicable. Because the aircraft is tracking
   along a magnetic course that must be in excess of 179 degrees
   (due to the crosswind), an altitude in excess of 3,000 feet AGL
   must be "even plus 500."

_______________________________________ A967

F. During the day and when less than 1,200 feet AGL, however,
   the pilot needs only a visibility of one mile and to remain
   clear of clouds.

Procedures & Regulations　　380

False. Leadville does not have an instrument approach. The highest airport with one is Telluride Regional Airport (also in Colorado). Field elevation there is 9,078 feet MSL, and the minimum descent altitudes for the three instrument approaches are 11,100, 11,600, and 12,380 feet MSL.

False. There is no minimum runway length specified for an instrument approach. The runway at Andover, New Jersey, for example, is served by a VOR approach and is only 1,981 feet long.

False. Although there are no regulations to this effect, airframe manufacturers usually do provide independent power sources.

D. The complete ground-to-air visual code is shown in Chapter 6 of the *Aeronautical Information Manual*.

False. According to 14 CFR §91.205, manifold pressure gauges are required only for "altitude engines," which are those equipped with turbo- or superchargers.

2255 and 2355 UTC. A telephonic search begins 30 minutes after the most current destination ETA. A decision to activate search-and-rescue efforts is made an hour later. Although en route position reports are not required when on a VFR flight plan, they are strongly encouraged.

A 974

48 hours (even if the battery has been used for up to 60 minutes
of operation and has been installed for half of its shelf life).

A 975

False. The tower controller will advise the pilot to switch to
ground control after he determines that the entire aircraft is clear
of the runway (beyond the runway holding position markings).

A 976

Climb, Communicate, Confess, Comply, and Conserve.

A 977

A displaced threshold elevates the glideslope (visual or electronic)
above any given point along final approach. This increases obstacle clearance and softens the noise footprint experienced by those
on the ground below approaching aircraft.

A 978

B. It is recommended also that a pilot maintain pattern altitude
    until passing the runway threshold when on the downwind leg.

A 979

False. The speed limit is 200 knots when within 4 miles of and
less than 2,500 feet above the primary airport even when operating outside Class D airspace. Refer to 14 CFR §91.117(b).

A 980

True. Although it might not be safe for a pilot not current at night
to land without the aid of twilight, 14 CFR §61.57(b) allows him
to do so while carrying passengers as long as he lands within an
hour after sunset (irrespective of how dark the sky might be).

A981

Illness, Medication, Stress, Alcohol (and drugs), Fatigue, and Emotional distress. It is recommended that each pilot preflight himself or herself for these factors prior to flight.

A982

D. A pilot is not required to file a report of such violation(s) unless requested to do so by the FAA. Refer to 14 CFR §91.3(c).

A983

True. FAA regulations require that a cross-country flight include a landing at other than the point of departure. *Voyager* pilots Richard Rutan and Jeana Yeager took off from Edwards Air Force Base in southern California on December 14, 1986 and landed there on December 23, 1986 without landing anywhere else during their globe-girdling voyage. Hopefully, they were granted an exemption. It would be shame to have lost 216 hours of cross-country flight time.

A984

United Airlines is operating its Flight 36 using a Boeing 747 with a Lufthansa Airlines' paint scheme (perhaps because United is leasing the aircraft from Lufthansa).

A985

A. Piston engines operated for hire must be overhauled upon reaching TBO, but those being operated privately need not be (as long as they remain airworthy). The TBO for turbine engines is mandatory whether operated for hire or not.

A986

C. At a groundspeed of 80 knots, a 12-degree descent profile would require a sink rate of 1,723 feet per minute (as compared to 425 fpm for a 3-degree glideslope).

C. According to the *Aeronautical Information Manual*, the prefix tango is used by air taxi or other commercial operators that do not have FAA-authorized call signs.

False. There are a number of such airports, including Kissimmee Municipal, Lake City Municipal, and Witham Field in Florida alone.

He must consult the applicable NOTAM. Because most altimeters cannot be set in excess of 31.00 inches Hg, the FAA will publish temporary flight procedures and restrictions for coping with this unusual condition.

Executive One.

This FAA-approved call sign is used by public-benefit flights such as those flown by Angel Flight and similar organizations.

False. Although most airports with runway lights do have rotating beacons, some do not, which can surprise those expecting to see a beacon in the distance when attempting to locate an airport at night.

False. There are many areas (particularly in the southwestern U.S.) where the floor of Class E airspace is several thousand feet above the surface. Refer, for example, to the Class E airspace south of Boulder, Nevada.

The "booze news" is how some pilots colloquially advise air traffic control that they have received ATIS Information Whiskey.

False. The statement is only partially correct. The aircraft also much be high enough to permit an emergency landing in the event of power failure without endangering people or property on the surface. Refer to 14 CFR §91.119(a).

Following an entire failure in 1956, the author (with arm extended) and his student, Griff Hoerner, were high enough to make a safe emergency landing on a sandy beach. The police officer, however, issued a cituation for illegal parking. (The ticket was thrown out of court.)   (A995)

$$A996$$

True. A pilot may log anything he'd like (including the time spent driving a Sherman tank). Such flight time, however, may not be used to satisfy currency or experience requirements.

$$A997$$

An operational ceiling is a maximum-allowable altitude. It most often is imposed because of a structural, aerodynamic, or system limitation (such as an oxygen system incapable of providing satisfactory sustenance above a certain altitude).

$$A998$$

True. Aircraft certification regulations allow up to a 10-percent overspeed under these conditions. This should be avoided, however, as propeller and engine damage could eventually occur.

$$A999$$

True. A glider with an engine is still a glider, not an airplane, and requires a glider rating. The amphibian may be flown as long as the pilot does not engage in water operations. (Flight through rain is allowed.)

$$A1,000$$

According to the *Outer Space Treaty of 1967*, airspace sovereignty ends where outer space begins. This altitude has yet to be legally fixed but varies between 85 and 100 NM. It most often is regarded as 90 NM (the altitude above which the control surfaces of an aircraft are considered ineffective).

$$A1,001$$

True. A test pilot must determine that no adverse effects (such as flutter) occur at what is called the dive speed (V_D). V_{NE} may not be greater than 90 percent of V_D.

Acknowledgments

Jay Apt
Robyn Astaire
Joseph Barber
Mark Bidgood
James Brous
John F. X. Brown
Larry Cagle
William Chester
Larry Clark
Jim Clarkson
Art Davis
John Deakin
Harv Denton
Scott Dickson
Michael Dolin
Don Draper
Christopher Eckland
Rob Edward
Ed Emanuel
Chris Fisher
John Fitzpatrick
Hal Fishman
Roger Fleishman
John Frank
John Friday
Jane Garvey
W. C. Garvey
Michael Gibbons
Harold D. Gordon
John V. Graff
James Griffin
David P. Hall
Phillip Hecksel
Yves Hoebeke
Tom Horne
Will Hubin
Chuck Jamieson
Drew Johnson

Ken Johnson
Larry Johnston
Phil Koch
Brian Korney
Raymond Maguire
David McGlumphy
Lonnie McLaughlin
Kathrynne McPherson
Brian Meyer
David Miller
Ward Miller
Steve Milliken
Phil Minschwaner
Jeff Mitchell
David Moss
Ben Moyle
Thomas Nagorski
Jack Norris
Frederick Pappas, Jr.
Jeff Pardo
Bruce Parker
Robert "Boom" Powell
John Price
Rob Sacks
Brian Schiff
Paul Schiff
Perry Schreffler
Dexter Senft
Irv Siegel
Matt Siegrist
John Sinnott
Richard Somers
Rob Spencer
James Terpstra
Anne Umphrey
Ed Williams
Penny Wilson